Propliners of the World
Volume 2

Front cover:

With a history dating back to the 1930s examples of the Consolidated PBY Catalina can be found in a number of guises. One is as an active water bomber and the other is, as illustrated, a flying yacht. Its ability to alight on either land or water make it it perfect aeroplane for the person who wants to get away from it all. This aircraft, N222FT is pictured at Oshkosh, Wisconsin.

Back cover clockwise from the top:

Developed for the early days of the US space program the Super Guppy still has a role to play with NASA to this day.

Turbine power has reached the only purpose built water bomber as this Canadair CL-415 runs in at Malta to drop its load of water.

Refurbished for a second life the Antonov An-2 awaits delivery to its new Russian owner.

About to depart to its summer fire base Grumman S-2T Turbo Tracker is one of a fleet operated by the California Department of Forestry.

'Tramp Steamer of the Sky' the Curtiss Commando is now operated only by a handful of companies in north and South America.

Propliners of the World
Volume 2

Gerry Manning

First published in Great Britain in 2011 by
Crécy Publishing

ISBN 9 780955 426858

Printed in Malta by Gutenberg Press Ltd

Crécy Publishing Limited
1a Ringway Trading Estate
Shadowmoss Road
Manchester M22 4LH

www.crecy.co.uk

CONTENTS

INTRODUCTION

In preparing this work the first task was to go through my slide collection to see what propliners I had photographed over the years. The number pulled out in the first instance was enough for several volumes, but on closer examination there were types no longer to be seen in the air, operators that had long ceased trading, and locations that no longer welcome nor host round-engine types. Sadly, the world of pounding pistons is in a decline and most of the operators are found in too few places. These classics have moved to some of the remotest places in the world, where they are simply irreplaceable.

Should the reader feel the urge to travel to Yellowknife, Fairbanks or Villavicencio, then do so because before too many years have passed the wonderful sound, smoke and smell of a piston engine bursting into life, on an operational aircraft, may have vanished for ever.

All the pictures in the book are my own from travels to many places over the years.

For those who would like to keep up to date with the subject, the excellent quarterly magazine *Propliner* is a must. For details visit www.propliner.co.uk

The basic premise of this book (in two volumes) is to look at some of the great classic propeller-driven aircraft as they operate around the world, sadly in ever-decreasing numbers. Each volume has four main sections.

In Volume 1 they are:

- The Douglas DC-3
- Bush and floatplane flying
- Preserved aircraft and pleasure flying
- First-generation turboprops

In Volume 2 they are:

- Water-bombers
- Russian and Chinese-built aircraft
- South American operations
- Cargo flights

Some aircraft, such as the DC-3, will appear in a number of sections. Of course the DC-3 deserves a section of its own, but also offers passenger flights, it still hauls cargo, and can be found in South America.

1 Water-bombers

The majority of the airlines of the world and the manufacturers of the latest aircraft try to say how 'green' they are with regard to the environment. They are, of course, right to show off this fact, but perhaps the greenest of all the aircraft in the world are small numbers of old piston-powered transport aircraft that run on gasoline, have round engines that belch smoke upon start-up and leave oil stains on the ramps from which they operate. How can this be? These aircraft are the 'water-bombers', whose sole task is to put out forest fires. People are often asked to print less to save paper and therefore a tree is saved for the good of the planet. What value must we therefore put upon an aeroplane that can save tens of thousands of acres of woodland?

The forest fire is a hazard of nature usually caused by a bolt of lightning striking tinder-dry areas, but humans, either by carelessness or malice, can also start a blaze. In parts of the world such as Canada the lumber industry is huge, and as well as the effects on nature the producers can see their raw material literally going up in smoke before their eyes.

The conventional fire engine cannot drive to the seat of the fire as there are no roads The blaze may also be on the side of a steep canyon – forests are not always flat areas. Therefore the only way to deal with such fires is from the air.

Over the last 50 years the aerial fire-fighting industry has grown from a few small crop-dusting aircraft spraying water on a local fire to government-run operations with high-technology equipment on station for the months of high risk, at vast expense, just in case they are needed. The cost of doing nothing is much higher.

In this work the term 'water-bomber' is used to cover the whole gamut of operations, although most aircraft do not use water but drop a line of chemical retardant that will block the fire's progress, and within the chemical compound is fertiliser to help generate fresh growth of trees. However, water is still used by the flying boats and amphibians, whose method of operation is to descend over a lake – forest areas often abound with them – lower some scoops under the fuselage and skim the lake to fill the aircraft's on-board tanks. This water is then dropped onto the fire from a series of doors in the tank, and the aircraft will keep repeating the process.

The types of aircraft used vary with who is operating them. Government operators, which usually have more money, will often run brand-new state-of-the-art aircraft designed for the task, while the private contractor may operate old yet well-maintained converted transports. Only in recent years have former Second World War bombers been withdrawn from the role, as in some cases owners were offered large sums of money for them from museums and warbird collectors, who continue to fly the aircraft but now in their old wartime colours.

The water-bomber scene is one that is always moving forward to protect the environment, and one of the most fascinating for the aircraft enthusiast.

A natural phenomenon, the forest fire is usually caused by a lightning strike. However, human behaviour is also often to blame, either from bits of discarded glass acting to magnify the rays of the sun on tinder-dry leaves and branches, by careless use of camping fires or, worst of all, by a deliberate act of arson. Forest fires can cover thousands of acres, and since natural forests do not have roads into them the fire engine cannot reach the seat of the inferno. This is where the water-bomber aircraft comes into its own, as it can fly over the fire and dump water onto it to extinguish it. The largest of all these prop-powered aircraft is the Martin Mars flying boat, which was designed as a transport for the US Navy and first flew in July 1942. Only six were built, and today just two still exist and have been saving the forests of British Columbia for many years. Pictured at its base at Sproat Lake, Port Alberni, Vancouver Island, is ***Martin JRM-3 Mars C-FLYL*** *(c/n 9267) 'Hawaii Mars', now operated by Coulson Flying Tankers.*

Top: *Everything about the Mars is big: it has a 200-foot (61-metre) wingspan, a length of 120 feet (36 metres), and can lift 7,200 US gallons (27,276 litres) of water for fire suppression. To do this the aircraft, a pure flying boat, not an amphibian, will lower scoops from beneath the fuselage and skim a lake at 70mph (112kmpr) for a total of 25 to 30 seconds to fill its tanks. To do this the lake needs to be between 4 and 5 miles long, and the aircraft needs to be able to clear an obstacle 50 feet high at either end of the run. Pictured at its Sproat Lake base is* **Martin JRM-3 Mars C-FLYK** *(c/n 9264) 'Philippine Mars'. Each of the six aircraft built were named after Pacific Ocean island groups.*

Bottom: *Both of the operational Mars are seen in this picture from Sproat Lake as 'Hawaii Mars' taxies past 'Philippine Mars' on its way to fight a fire. The normal crew for missions is two pilots and two flight engineers, and all throttle movements are made by the engineers on instructions from the pilots. It is only on the lake while taxiing that the pilot operates the power settings, and in the final seconds of a landing or a drop. The inboard engines have a reverse propeller pitch for water manoeuvring. Only high-hour pilots with lots of time on water are allowed to fly the aircraft, as the drop height is as low as 150-200 feet (45-60 metres). A single drop can cover an area of 3-4 acres (1.2-1.6 hectares), and each of the aircraft can drop, skim and drop again at approximately 15-minute intervals.*

Top: *To bring a Mars in from the lake for major maintenance, beaching gear has to be fitted while the aircraft is in the water; only then can it be pulled up the ramp. The scale of the aircraft is apparent in this picture of* **Martin JRM-3 Mars C-FLYK** *(c/n 9264) at its Sproat Lake base. It is usually only at the end of the fire risk season in autumn and winter that the aircraft will exit the water. The aircraft have proved to be so useful that one of them can often be found operating from Lake Elsinore in California. Almost all the operations for the Mars are from freshwater lakes, so saltwater corrosion is not a problem.*

Bottom: *This close-up of* **Martin JRM-3 Mars C-FLYK** *(c/n 9264) out of the water at Sproat Lake shows the size of the engine. The powerplants for this aircraft are four 2,500hp Wright Cyclone R-3350 air-cooled radial pistons, now running on 100-octane fuel, and each has an oil capacity of 90 US gallons (353 litres). With an endurance of 6 hours per flight, the Mars can help the Canadian timber industry from losing its raw material in the form of burning trees. During recent years, before the current owners Coulson Flying Tankers took over the operation, it looked as though the pair might be retired to museums but, since no other aircraft can perform its role, the 'Super Scooper' will hopefully fly on for many years to come.*

EC-EVK

C-FPQM
714
Buffalo AIR
NORTHWEST TERRITORIES, CANADA

714
703
Buffalo AIR

Top left: *Another aircraft that scoops water from lakes is the Consolidated PBY Catalina. This had the largest production run of any flying boat, with more than 3,000 airframes constructed. It was an amphibian, so could operate from both water and airfields. Pictured at Madrid's Cuatro Vientos Airport is* **PBY-6A Catalina EC-EVK** *(c/n 2008), operated by SAESA (Servicios Aereos Espanoles), a Spanish aerial fire-fighting company. The European water-bomber operation is not as large as those of the USA or Canada, but the hot southern countries bordering the Mediterranean have a risk of brush and grass fires that can threaten centres of population as well as the large woodland areas.*

Middle left: *Although in declining numbers, the PBY still has a role to play in fire-fighting, and one of the most important is that of cost. The airframes all date back to the Second World War and are relatively cheap to acquire, and if they have been with a company for a number of years, annual depreciation reduces the book value to low figures. When the company is bidding for a contract, all it needs to calculate is fuel, maintenance and crew costs. If the summer is quite wet and there are few fires, the aeroplane does not fly very often, and to have a brand-new very expensive aircraft do so little flying does not make good economic sense. The biggest threat to PBY operations is the demand for the type from the warbird community or from museums, which are prepared to pay large sums for the type and preserve them in a different role. It is of note that such types as the Boeing B-17 Flying Fortress and the Grumman F7F Tigercat could be found in water-bombing operations until their value grew too high to refuse offers from warbird collectors. Pictured at Red Deer, Alberta, is* **Consolidated PBY-5A Canso C-FPQM** *(c/n CV-425), operated by Yellowknife-based Buffalo Airways. The Canso is the name given to PBYs built under licence in Canada.*

Bottom left: *This view shows two of Buffalo Airways' PBYs at the company's southern store and maintenance base at Red Deer, Alberta. The first flight of the Catalina was as far back as March 1935, from Naval Air Station Anacostia, Washington DC. It had a long service career and during the second World War its long range was vital in patrols over the oceans of the world. Single actions included finding the German battleship Bismarck after it had sunk the Royal Navy's HMS Hood, as well as the destruction of many U-boats. Powerplants were a pair of Pratt & Whitney Wasp R-1830 fourteen-cylinder radial pistons with an output of 1,200hp.*

Below: *The need for a fire-fighting aeroplane to protect its vast lumber industry saw Canada become the first nation to design a purpose-built water-bomber aircraft. Following wide-ranging consultations, a twin-engine amphibian was decided upon, with the ability to scoop and also be loaded at a base; moreover, it would have tanks with multiple drop doors to vary the amount that fell in one pass over the fire. Many hours of work went into such features as the shape of the hull, the size of the scoops and the dump doors. The first flight of the design, the Canadair CL-215, took place in October 1967 as a landplane, with the first water take-off in May of the following year. Power came from a pair of Pratt & Whitney R-2800 eighteen-cylinder air-cooled radial pistons with an output of 2,100hp. Pictured at Abbotsford, British Columbia, is* **Canadair CL-215 C-GFSM** *(c/n 1098), owned by the Province of Alberta but on lease to locally based Conair.*

Top: *Red Deer, Alberta, is the location of this* ***Canadair CL-215 C-GFSK*** *(c/n 1085), owned by the Province of Alberta. It the early days of operating the type it was always the provincial governments that owned and operated them as the cost of purchase for a private operator, and flying less than 200 hours all year, would not be economic. Despite the high costs involved in the operation of a fleet of water-bombers, they can quite often pay for themselves as the cost of the timber in the forests they save is more than the cost of operation of the aircraft. They operate from a number of airfields across the province, and during the high-risk summer months a pair of aircraft will be based and be on call for the daylight hours seven days a week. As well as the big Canadair, a small twin-engine aircraft is also based, and has the role of spotter and guide for the water-bomber with regard to the best approach line to deal with the fire.*

Bottom: *The Province of Ontario has large forest regions, with 66% of the state covered in trees. These comprise spruce in the north, with maple and oak in the south, together making an estimated 85 billion trees in all. Not all fires are bad for forests, as they burn off dead leaves and underbrush and create a clear area for either nature or lumber companies to reforest. To do this control burns take place, which need to be under strict supervision. Ontario's Aviation & Forest Fire Management has been based at Sault Ste Marie since 1924. Pictured at the Dryden Airport base is* ***Canadair CL-215 C-GENU*** *(c/n 1082), then operated by the province. It has both Pratt & Whitney engines turning as it runs a powerplant test.*

Top: *Pictured waiting for the fire alarm to sound are a pair of Ontario province-owned CL-215s at their Dryden base. The type is renowned for its high productivity; during a 1-hour period, one aircraft operated in Quebec made thirty-one pick-ups and drops, while in the former Yugoslavia one aircraft dropped 300,000 gallons of water during 225 pick-ups and drops. Fire-fighting has some high-tech help; the Dryden base even has a lightning counter, which marks a map with all the strikes. Operations staff can therefore see the locations, and if there are a lot in one area they may assume a high risk of an outbreak there.*

Bottom: *Winnipeg is the home base of the CL-215s operated by the Province of Manitoba. Pictured on the ramp at base is* ***Canadair CL-215 C-GMAK*** *(c/n 1107). All the aircraft used in fire-suppression operations carry a three-number code, in this case No 254. For many piston-powered aircraft the need to either re-engine them with turboprops or replace them with new-built CL-415s becomes more and more urgent as the price and, more importantly, the availability of AVGAS (aviation gasoline) reaches unacceptable levels. The province plans to replace the piston-engined versions over the next few years.*

250
Manitoba
251
Manitoba

1039

Top left: *Three of the Province of Manitoba's fleet of Canadair CL-215s are pictured at Winnipeg in early autumn. Once the season for fires is over and the rain has started to fall, the fleet will return to its main base from its outstations across the province for maintenance and storage during the long, cold Canadian winter.*

Bottom left: *As well as Canadian operators, the CL-215 has been sold to a number of governments and air forces around the world, including Croatia, France, Italy, Portugal, Spain and Turkey. In Venezuela they have a dual use, as both a water-bomber and a transport aircraft seating up to twenty-six passengers. In Thailand the type is used for maritime patrol with fire-fighting as a secondary role. Pictured landing at Athens is Greek Air Force* ***Canadair CL-215 1039*** *(c/n 1039), operated by Elefsis-based 355 Mira (Squadron). Despite its military operation the aircraft has retained the almost standard bright yellow livery that makes it highly visible over a forest area or in the thick black smoke of a large fire.*

Top: *The Province of Saskatchewan operates its CL-215s in a basic white colour with green trim rather than the usual yellow. Pictured at the air tankers' main operating base at La Ronge is* ***Canadair CL-215 C-FAFO*** *(c/n 1094).*

Bottom : *Winter storage awaits this line of four CL-215s at La Ronge, Saskatchewan.*

Top: *As some of the provincial governments of Canada replace their older CL-215s with new-built turbine-powered models, the piston-powered aircraft are finding their way to some of the civilian operators. Pictured at Red Deer, Alberta, is* ***Canadair CL-215 C-GDHN*** *(c/n 1089) with Buffalo Airways titles; the Yellowknife-based carrier has leased four airframes from the Northwest Territories Forest Management Division. The Government's Department of Aviation services operate six bases across the vast area of the territory, at Fort Smith, Hay River, Fort Simpson, Yellowknife, Norman Wells and Inuvik. They also have reciprocal agreements with Alberta, Yukon and British Columbia to help each other out should a very large blaze overwhelm the local unit and extra aircraft and pilots are needed.*

Bottom: *Pictured at the company base of Red Deer, Alberta, before its sale to Turkey, is* ***Canadair CL-215 C-FTUU*** *(c/n 1011) operated by Air Spray, a company that has been involved in aerial fire-fighting since 1967. As well as actual operations, it has provided a full maintenance service for its fleet, as well as the conversions of standard aircraft with tanks and drop systems fitted for their new roles.*

Top: *The Air Spray name is clearly painted upon these two CL-215s at Red Deer prior to their sale to Turkey. The company had operated the type since 1986 from the many lakes in Alberta.*

Bottom: *Pictured at Red Deer, Alberta, undergoing maintenance, is* ***Canadair CL-215 C-GORF*** *(c/n 1104) operated by locally based Air Spray. It is apparent how large the type is when compared to the steps at the door. The aircraft has a wingspan of 93ft 10in (28.6 metres), a length of 65 feet (19.82 metres), and a height of 29ft 6in (8.98 metres). It can cruise at 181mph (157 knots/291kmph) with a range of 1,405 miles (2,260km) flown by a crew of two.*

Below: *It was logical development to fit turboprops to the CL-215, and the new powerplants were a pair of Pratt & Whitney Canada PW-123AF engines; when retro-fitted, the designation CL-215T was used. However, the manufacturer, Canadair, now part of the Bombardier group, decided to build a completely new variant, the 415. This first flew in December 1993 and operates in the same manner as its predecessor as a scooping amphibian. It can upload 1,621 US gallons (6,137 litres) of water in 12 seconds and has various doors on the tanks enabling it to drop the water either all at once or in different patterns using different door combinations. Pictured at Luqa, Malta, is* ***Canadair CL-415 I-DPCF*** *(c/n 2059) operated by the Italian Protezione Civile. It shows in both the frontal and side view the changes in the new aircraft. Note that the elevators on the tail and the wing tips have extra additions to them.*

In Italy the CL-415 aircraft are operated on behalf of the Protezione Civile by SOREM (Societa Richerche Esperienze Meteorologiche Srl), based at Rome-Ciampino with a fleet of nineteen as well as five-piston powered CL-215s. ***Canadair CL-415 I-DPCF*** *(c/n 2059) is pictured on approach, then dumping water at Luqa, Malta. Popular in the Mediterranean area, the type can drop 14,600 US gallons (55,260 litres) of water or foam mix within the space of an hour from a water source 6 nautical miles (11km) from the location of the fire.*

Below: *In France the Sécurité Civile based at Nimes-Garons operates the state-owned water-bombers. This organisation was one of the launch customers for the original piston-powered CL-215s. It has sites all around the French coast, with the main water-bomber operations at fire bases along the Mediterranean coast together with two more on the French island of Corsica. Pictured at Cazaux in south-west France is* **Canadair CL-415 F-ZBFY** *(c/n 2010) of the Sécurité Civile. French pilots have fewer lakes than their Canadian counterparts and have to undertake open-sea skimming, where there may be choppy waves. To train for full operations pilots will always begin on calm lakes before open-sea scooping, and then always under the instruction of an experienced captain.*

Top right: *Conair, based at Abbotsford, British Columbia, is one of the leading aerial fire-fighting companies in the world and has more than forty years of experience. The company is also the manufacturer of one of the most widely used types, the Conair Firecat. This aircraft started its life as the Grumman S-2 Tracker, with a role as a carrier-borne anti-submarine aircraft for the US Navy. It first flew in December 1952, followed by a production run of more than 1,200 airframes, including 100 aircraft built under licence by de Havilland of Canada for the Royal Canadian Navy. It also served in the armed forces of Argentina, Australia, Brazil, Italy, Japan, Korea, Holland, Peru, Taiwan, Turkey, Uruguay and Venezuela; many of these nations did not have aircraft carriers, so operated them from shore bases. The powerplants were a pair of Wright R-1820 air-cooled radial pistons with an output of 1,525hp. Pictured here heading the line, at base, is* **Conair Firecat C-FOPY** *(c/n 019), one of the licence-built Canadian examples and operated by the company itself.*

Middle right: *Starting life with the Canadian Navy before its conversion, in March 1983, is* **Conair Firecat C-GWUP** *(c/n 012) in the company colours at its Abbotsford base. The conversion to its new role will involve a major rebuild: the weapons bay will be removed and in its place will be fitted a 725-imperial-gallon (3,296-litre) tank for the fire-fighting retardant or water, which will fit within the fuselage and will not bulge out under it. There are four doors in the tank and various drop sequences can be performed as required by the circumstances of each fire. Aircraft are flown by a single pilot and the type has a fatigue life of 12,000 hours.*

Bottom right: *Pictured at its Abbotsford base is* **Conair Firecat C-FEFK** *(c/n 014). Unlike most of the conversions, this was a Grumman-built example that served in the US Navy. Conair acquired the aircraft in 1982 and converted it the following year. It then left for a six-year stint with France's Sécurité Civile before returning to operate with Conair in Canada. Another of the modifications during the conversion is to increase the size of the main wheels, as the originals were not always suited to some of the more remote operating sites when the fleet is dispersed during the summer operations. To facilitate this the size of the undercarriage doors had to be increased.*

CONAIR FIRECAT
C-GWUP

74
C-FEFK
CONAIR FIRECAT

571

571

CONAIR FIRECAT

Left page: *When autumn arrives, the fleet will return to Abbotsford for winter storage and maintenance. These three views show the Conair Firecats parked up with blanking discs over the front of the engines. A rain-filled sky is about to unleash itself upon the aircraft.*

Top: *What might seem to be wrecks are in fact ex-military S-2 Trackers in store for either conversion to a Firecat, if Conair get an order for one, or for spare-part reclamation to keep the existing fleet in the air. This picture, at Abbotsford, shows seven examples, part of what was one of two lines of stored airframes.*

Bottom: *The Province of Saskatchewan's aerial fire-fighting headquarters is in the north at La Ronge. Pictured in the hangar, at base, is 1960 Canadian-built* ***CS2F-2 Tracker C-GEQE*** *(c/n DHC-92). The province has its own white and green livery on the aircraft as it begins its end-of-season maintenance. The wings are removed for non-destructive testing, the undercarriage is overhauled and the engines are removed for service. The La Ronge facility also has a certified avionics workshop as well as a spares holding of several million dollars worth of units, many of which are no longer in production. During the summer the province has another six operational bases as well as La Ronge, at Buffalo Narrows, Prince Albert, Meadow Lake, Hudson Bay, Baker Narrows and Stony Rapids. If they are at low risk and other provinces have a large fire, they may be asked to deploy and assist; they have been as far east as Newfoundland and west to British Columbia.*

503

503
C-GEHR

Left and Above: *Saskatchewan's Canadian-built* ***CS2F-2 Tracker C-GEHR*** *(c/n DHC-51) is pictured at its La Ronge base running its two Curtiss-Wright R-1820 1,480hp air-cooled radial piston powerplants. It has loaded its tanks with water for a practice flight. In the second shot the aircraft is taking off, in the third it makes a low run over the trees, and in the fourth it drops its load. The aircraft will usually drop from an altitude of about 140 feet, often just 30 feet above tree height, and speed will be in the region of 105 knots. The role of the aircraft is not always to drop the retardant on the seat of the blaze but to plant it in the direction the fire is moving to stop it advancing. This can create a safe area for the fire to burn itself out.*

Below: *As well as being one of the early customers of the Firecat, France's Sécurité Civile has had its aircraft returned to Abbotsford for Conair to fit turbine engines to them; all their aircraft are now turboprop-powered. Pictured at Cazaux in south-west France is* ***Conair Turbo Firecat F-ZBAA*** *(c/n 027) with its Pratt & Whitney PT-6A turbines turning. This aircraft was built by Grumman for the US Navy, and when it was demobbed Conair converted it to a piston-powered Firecat in 1987. It then flew to France for operations with the Sécurité Civile. Five years later, in 1992, it flew back to Canada's west coast for conversion by Conair to turbine power. This was completed the following year and the long flight back to its French home followed.*

Top: *It is a logical question to ask, why use smaller aircraft with limited loads when larger types with huge capacity are available? The answer is quite simple: smaller aircraft are usually faster and therefore get to a fire more quickly, hopefully to nip it in the bud before the big aircraft arrive upon the scene. That is why the California Department of Forestry and Fire Protection (CDF) uses a fleet of twenty-three ex-US Navy Grumman-built S-2 Trackers. They were acquired in 1996 from military-surplus stocks and over the next nine years were converted to S-2Ts with a 1,200-US gallon tank and a pair of Garrett TPE-331 turbine engines with a power output of 1,650shp each. Pictured at its Sacramento base, the former McClellan AFB, is* ***Grumman S-2T Tracker N431DF*** *(c/n 109C), showing off its new powerplants.*

Below: *The CDF has house colours of white aircraft with red trim, and the tanker's individual number is on the tail, as are the 'CDF' initials of the fin tip. The aircraft is flown by a single crewman and has an endurance of 4.5 hours with a loaded range of 500 miles, 800 when empty. Pictured at its Sacramento base is* ***Grumman S-2T Tracker N450DF*** *(c/n 228C), Tanker No 93. It was in the 1950s that the CDF first started to tackle fires from the air, when agricultural crop-sprayers were used to drop water. California, with its mix of forest, crops and prime housing, has the largest operation of any state to combat the menace of fire.*

Top right: *Pictured at the Sacramento base is a line of six Grumman S-2T Trackers. They are about to depart to their summer bases throughout the state of California. The leading aircraft has, above the CDF badge on the nose, its base location, Hollister, one of thirteen operated during the fire season, which runs as late as October.*

Above: *Pictured at its Ramona Air Attack base is* ***Grumman S-2T Tracker N431DF*** *(c/n 109C), Tanker No 78, and its 'bird-dog'* ***North American-Rockwell OV-10A Bronco N409DF*** *(c/n 305-18M-12). This latter type was designed for counter-insurgency and close air support for ground-based troops, and carried a crew of two. Its new role, having been demobbed from US military service and civilianised, has been to act as a flying command and control aircraft during fire-fighting operations. The crew of two today comprises a pilot and an Air Tactical Group Supervisor. This person will guide the tankers in and tell them where to drop their retardant or water to stop a fire spreading. If it is a very large fire – they can cover hundreds of thousands of acres – and fleets of tankers are tackling it, they will also act as a flying air traffic control to keep the tankers in the order they want them to attack the blaze and prevent collisions in what will also be an area covered in smoke.*

2

C-FOVC
56
1
13
14
20
AIR SPR
56

98

Top left: *Pictured at its Red Deer base is* **Douglas A-26C Invader C-FPGP** *(c/n 29177) of Air Spray with both its Pratt & Whitney R-2800 air-cooled radial pistons turning as it tests its oil consumption.*

Middle left: *The Douglas A-26 Invader has had a long military history as well as a water-bombing one. It first flew in July 1942 and following Second World War service found roles in Korea, Vietnam, with the French in Indo-China, with the CIA in the 1962 Bay of Pigs operation in Cuba, and in Nigeria's civil war with the breakaway province of Biafra. It was designated 'A' for attack with the USAAF, but in 1948 the now USAF changed this to 'B' for bomber. The original B-26 had been the Martin Marauder, but was by then long out of service. However, the 'A' designation has stuck with a lot of operators. The type was also the basis for a number of post-war executive transports, with different companies converting aircraft to a new civil role. The lift capability, speed and manoeuvrability of the A-26 made it a natural for conversion of many of the available airframes to the water-bomber role. The largest user of the type was Air Spray of Red Deer, Alberta, which at one time operated eighteen; they are now in store at base. Pictured leading the line is* **Douglas A-26C Invader C-FOVC** *(c/n 28776).*

Bottom left: *Pictured at Red Deer, when in service, are three Air Spray Douglas A-26 Invaders awaiting the start of their summer duties. The company is contracted by the Alberta Forest Protection Division to provide a fixed number of aircraft at a set location for a time period of up to 100 days. If this starts at the beginning of June it will run into September and the start of the low-risk period. Contracts would of course be extended if fires were still a high risk. Ten bases are in operation, from the south of the province at Pincher Creek to the far north at High Level. As well as the tanker aircraft and pilots, a 'bird-dog' spotter and pilot join the base with a small team of ground engineers to keep the aircraft in peak condition. Two states of alert are kept, Red and Yellow; in the former, crews are on site and ready to fly with the 'bird-dog' airborne within 5 minutes, while in state Yellow the 'bird-dog' has 30 minutes to take off while the tankers are loaded.*

Below: *Another Second World War aircraft that has recently retired from water-bomber operations but is preserved as a flying warbird is the Consolidated PB4Y-2 Privateer. It was developed from the B-24 Liberator, and its role was to be that of a maritime patrol aircraft for the US Navy. The biggest difference between the two types was the replacement of the twin fins with one tall single unit. Other changes were fuselage length and the shape of the engine nacelles. The powerplants were four Pratt & Whitney R-1830 air-cooled radial pistons with an output of 1,350hp. These did not have the turbo-superchargers fitted to the B-24, as high-altitude operations were not required for maritime patrol. The type saw service in Korea in the role of flare-dropping over road convoys, thus illuminating the scene for US Marine Corps attack aircraft. The main water-bomber operation was with Hawkins & Powers of Greybull, Wyoming, which fitted the aircraft with a bomb-bay tank with a capacity of 2,400 US gallons and replaced the engines with Wright Cyclone R-2600s having an output of 1,700hp each. This gave the aircraft a better performance in hot and high ground areas. Pictured at Lancaster, California, is* **Consolidated PB4Y-2 Privateer N6884C** *(c/n 59701), then operated by Hawkins & Powers.*

Top: *All three of the four-engine piston-powered Douglas airliners have found a role in fire-fighting. First of the three was the DC-4, which took to the air for the first time in February 1942 from Santa Monica, California. Wartime needs saw it in a transport role as the C-54. As the type was retired from passenger service, latterly with holiday charter operators, it was soon being used for cargo-hauling and also fire suppression. The aircraft have a tank fitted under the fuselage with a capacity of 2,000 US gallons. Pictured on fire base duty at Chico, California, is* **Douglas DC-4/C-54 N82FA** *(c/n 35960) of Aero Flite, based in Kingman, Arizona. This aircraft was built for the US military in 1945 and served until 1973, when it was stored before taking civil markings four years later. Aero Flite currently operates the Canadian CL-215.*

Bottom: *Pictured at Dinuba-Sequoia airfield, California, is* **Douglas DC-4/C-54 N8502R** *(c/n 27367), Tanker No 65, operated by TBM Inc. This company has been in the aerial fire-fighting business since 1957, and its name derives from the Grumman TBM Avengers it first operated. The company has also flown such Second World War types as the Boeing B-17 Flying Fortress and the Grumman F7F Tigercat, both of which types have been snapped up by the warbird movement. The pictured aircraft was delivered to the USAAF in 1945 and served until 1971, when it went into store. In 1978 it was sold and in the years that have followed has had a number of different owners and operators.*

Top right: *Once owned by Central Air Services,* **Douglas DC-4/C-54 N67040** *(c/n 27232), Tanker No 147, has been in store at Avra Valley, Tucson, Arizona, for at least ten years. However, the location and climate are perfect for long-term storage, as it is hot with low rainfall and humidity. It is of note that the main storage base for the US military, Davis-Monthan AFB, is located just a few miles away. Built in 1944, this aircraft served with the US Navy until 1967, when it moved to Davis-Monthan for storage. It left in 1975, having been sold to a civil operator, and over the following years has had a number of owners.*

Middle and Bottom right: *Pictured on water-bomber duty at Lancaster, California, is* **Douglas DC-4/C-54 N9015Q** *(c/n 22178), Tanker No 152, operated by ARDCO Inc (Aerial Retardant Delivery Company), based at Ryan Field, Tucson, Arizona. This airframe is a 1945-built one that served with the US military until 1978, when it arrived at Davis-Monthan for a period in store. This was short-lived, as later the same year it entered the world of civil operators with the first of many owners and operators. The second picture shows the aircraft taking off at Santa Barbara, California, a few days later, when it had called in for fuel and a tank full of retardant while fighting a local fire. This burn had reached a size that brought aircraft from other bases to deal with it.*

147
150

ARDCO Inc.
N9015Q

152
N9015Q
ARDCO Inc.

Top: *With its annual season as a fire-bomber over,* **Douglas DC-4/C-54 C-FBAJ** *(c/n 3088), Tanker No 02, is pictured at Hay River, Northwest Territories. It is owned and operated by Yellowknife-based Buffalo Airways, which, as well as water-bombing, operates cargo and passenger flights in piston-powered aircraft. This aircraft was built in 1943 and was operated by the US military until the war's end; it was then taken over by the federal government for the next sixteen years, when it was sold on. It was converted to water-bomber status by Aero Union, based at Chico, California, which also manufactures the tank fitted under the fuselage. It came to Canada in 2000 and now wears Buffalo's green livery.*

Middle: **Douglas DC-4/C-54 C-FBAP** *(c/n 36089), Tanker No 15, of Buffalo Airways, still retains the attractive colour scheme of its previous owner, Aero Union. It now has the Buffalo name on the cabin roof as well as its type name for it, 'Fireliner'. This particular aircraft was built in 1945, served with the US military until 1973 and, following five years in store at Davis-Monthan AFB, was sold on the civil market. It was bought by Aero Union in 1982, which converted it to a water-bomber, later selling it to Buffalo.*

Bottom left: *The next of the big Douglas pistons was the DC-6, which first took to the air in February 1946 from Santa Monica, California, and can still be found in service in Alaska with a number of cargo and fuel carriers. When converted to a water-bomber, the tank capacity varied from 2,400 to 3,000 US gallons depending upon the company undertaking the conversion. The type was powered by four Pratt & Whitney R-2800 air-cooled radial pistons with an output of 2,500hp each. Pictured at its Dinuba-Sequoia, California, base is* ***Douglas DC-6A N90739*** *(c/n 43044), Tanker No 68, of TBM Inc. The DC-6 looked very similar to the DC-4 externally; the biggest change was the introduction of a pressurised cabin. The DC-4 was 93ft 10in (28.6m) long with a wingspan of 117ft 6in (35.81m), whilst the DC-6 had a length of 100ft 7in (30.66m) for the original and 105ft 7in (32.18m) for the A and B models, with the same wingspan of 117ft 6in as the DC-4 on all versions.*

Top: *Conair, based at Abbotsford, British Columbia, has operated the DC-6 for many years going back to the 1970s. Pictured at base is* ***Douglas DC-6A C-GKUG*** *(c/n 45177), Tanker No 50. This aircraft was built in 1957 for Western Airlines, based at Los Angeles. The carrier had a history dating back to 1925, but was taken over by Atlanta-based Delta Air Lines in 1987. Following this the aircraft joined the Chilean Air Force from 1966 to 1982, when it moved to its present owner, Conair.*

Bottom: *Pictured at Abbotsford, BC, is Conair's* ***Douglas DC-6A/C C-GIBS*** *(c/n 45531), Tanker No 51, parked out on the grass for winter storage. Built in 1958 for British airline Hunting-Clan, it later moved to Switzerland for several years flying holidaymakers to the sunspots of Europe before joining the Conair fleet of water-bomber conversions in 1982.*

Top: **Douglas DC-6B C-GHLY** *(c/n 45501), Tanker No 46, is one of the oldest aircraft in the Conair fleet. Pictured at its Abbotsford base, it was built for Northwest Orient Airlines in 1958, a Minneapolis-based carrier that has in recent times been taken over by Atlanta-based Delta Air Lines. Following service with Northwest it journeyed to Finland in 1965, then to Belgium in 1972. Two years later it was bought by Conair for conversion to its role as a water-bomber.*

Bottom: *The last of the Douglas pistons was the DC-7. It was also the first of the big piston line to be a water-bomber, albeit by accident. It first flew in May 1953 and later that year, while on a test flight over Palm Springs Airport, the crew dumped a load of 1,300 US gallons of water ballast. It produced a path nearly a mile long and 200 feet wide, which prompted the Los Angeles County Fire Department to run some tests using the aircraft later that year over Rosamund Dry Lake. A load of 2,400 US gallons of water, divided into six tanks feeding nozzles for the dumping, was lifted into the high desert of California. All proved very satisfactory, but nothing happened due to costs. At that time the DC-7 was the very pinnacle of piston-powered aircraft design and simply too expensive to be operated for the low number of hours per year that is normal for water-bombers. Moving on twenty years, the DC-7s were looking for buyers lest they be scrapped, and they found their new role, albeit now with proper belly tanks. Pictured at Palmer, Alaska, is* **Douglas DC-7B N4887C** *(c/n 45351), Tanker No 33. It is owned by International Air Response, formally known as T&G Aviation, based at Coolidge Municipal, Arizona.*

Top: *The biggest of the land-based piston-powered water-bombers has been the Boeing C-97. First flown in November 1944, the type had two versions: a transport with the US military, and the Stratocruiser, in its day the most luxurious airliner flying. The military C-97 was in service with the USAF and Air National Guard until 1977 in the role of an airborne tanker to refuel jet fighters and bombers. It was powered by four Pratt & Whitney R-4360 radials with an output of 3,500hp each. Its wingspan was 141ft 3in and its length 110ft 4in. Only one was converted to water-bomber operations, and was fitted with a 3,000-US gallon tank. Pictured upon its arrival for the summer season at Fort Wainwright, Fairbanks, Alaska, is* **Boeing C-97G N1356N** *(c/n 16729), Tanker No 97, then owned by Hawkins & Powers of Greybull, Wyoming.*

Bottom: *The Lockheed Neptune was the mainstay of the US Navy's maritime patrol squadrons from 1947 until the mid-1960s with reserve units. It first flew in May 1945, and in September 1946 one famous aircraft, named 'The Truculent Turtle', flew non-stop from RAAF station Pearce near Perth, Australia, to Columbus, Ohio. It had covered 11,236 miles in 55hr 17min, a record that stood for many years. In the late 1970s the US Forestry Service developed the Neptune as a water-bomber. Two main companies have done the conversions and in many ways they are quite different. Pictured at its Chico, California, base is* **Lockheed SP-2H Neptune N718AU** *(c/n 726-7214), Tanker No 18, of Aero Union. Note that this aircraft has a smooth underbelly with the tanks inside the fuselage; the auxiliary jets for boosting power on take-off have been removed, yet the MAD (Magnetic Anomaly Detector) extended tail boom has been retained following the conversion by Aero Union.*

Top: *This Neptune, operated by aptly named Neptune Aviation of Missoula, Montana, is one of the conversions pioneered by Black Hills Aviation of Alamogordo, New Mexico. This company, now taken over by Neptune, had the 2,700-US gallon tank, with six doors, visible under the fuselage. The MAD tail boom has been removed and the 3,400lb-thrust Westinghouse J-34 turbojet under the wings has been retained. Pictured on duty at Lancaster, California, is* ***Lockheed P2V-7 Neptune N443NA*** *(c/n 726-7168), Tanker No 43. The company operates a fleet of ten aircraft, and as well as the main Montana base has retained the former Black Hills base in New Mexico.*

Bottom: *Pictured at Avra Valley, Arizona, is* ***Lockheed P2V-7 Neptune N229MA*** *(c/n 726-7211), Tanker No 99 of Minden Air Corp based at Minden, Nevada. This company was established in 1990 and plans to be the first operator of the BAe-146 jet airliner in aerial tanking operations.*

Top right: *The Neptune is powered by a pair of Wright R-3350 air-cooled radial pistons with an output of 3,500hp each. Pictured at Lancaster, California, is* ***Lockheed P2V-7 Neptune N14447*** *(c/n 826-8010), Tanker No 11 of Neptune Aviation, with both engines turning the four-bladed propellers. If there are no fires to put out, the aircraft must have their engines run on a regular basis to ensure that when they are needed they will start first time. Pilots also have to keep their experience up to date and will fly scheduled test drops if there are no actual fires to tackle.*

Middle and Bottom right: *This pair of views shows three of the five Neptune water-bombers to be seen at Lancaster on that day. This was a higher than usual number due to the proximity of a large fire that had been brought under control. In fact, just a few days earlier eight aircraft were on base operating a continuous shuttle to dump retardant on the blaze.*

NEPTUNE AVIATION SERVICES, INC.
11
MISSOULA, MT

NEPTUNE AVIATION SERVICES, INC.
05
MISSOULA, MT
N4235T

MONTANA
05
09
N4235T
43

Top: *One of the newest conversions of a type to water-bomber status is that of the Convair 580. The Convair range of 240/340/440 models are 1950s airliners, all piston-powered, and during the following decade many had the original engines replaced by turboprops to give the aircraft an extended life in both passenger and cargo roles. The water-bomber conversions are a twofold exercise between a pair of companies, both based in British Columbia. The first operation is by Kelowna Flightcraft, which, as stated earlier, now holds all the design authority for the type. Following that company's work the aircraft will fly across the province to Abbotsford to be fitted with the retardant tank, designed and built by Conair. It is then ready for operations. Pictured at its Abbotsford base is* ***Convair 580 C-FHKF*** *(c/n 374), Tanker No 55, owned and operated by Conair.*

Bottom: *Powerplants for the 580 are a pair of Allison 501 turboprops with an output of 3,750shp each. This gave the re-engined machines much more power following the conversion. Pictured at its Abbotsford base is Conair-owned* ***Convair 580 C-FKFA*** *(c/n 100), Tanker No 52, with both engines running at full power during a test. This aircraft was built in 1953 as a model 340 for Atlanta-based Delta Air Lines and re-engined to bring it up to 580 standard in May 1969. Over the following years it has had many owners and operators. It was one of two that, in December 2010, flew to Australia to operate for the State of Victoria for a three-month period during the southern hemisphere summer. The ferry flight took 38 flying hours, the route being Abbotsford to Oakland, California, then on to Honolulu, Majuro, Cairns in Queensland and down the coast to Avalon.*

Top: *From Air Force One to a water-bomber has been the fate of this aircraft, now owned by the Province of Saskatchewan and pictured at its La Ronge headquarters.* ***Convair 580 C-GSKQ*** *(c/n 217), Tanker No 475, was built in 1954 as a model 340 for General Dynamics Corp, but instead joined the USAF under the designation C-131D. In 1966 it had its original engines replaced with turboprops and was re-designated VC-131H (the 'V' was for VIP). It was in October 1972 that it flew President Richard Nixon to Huntington, West Virginia, an airport too small for the usual jets. Any aircraft that has the US President on board will have the right to use the call sign 'Air Force One'. As well as the President, it frequently flew his deputy, Vice-President Gerald Ford, the call sign then being 'Air Force Two'.*

Middle: ***Convair 580 C-GSKR*** *(c/n 509), Tanker No 471, operated by the Province of Saskatchewan, is pictured at its La Ronge base. This view shows the size of the huge 13ft 6in Aeroproducts A644 propellers. It also shows off the underbelly retardant tank, with its capacity of 2,100 US gallons (7,950 litres).*

Bottom: *A view of three of the Province of Saskatchewan's Convair 580s lined up at their La Ronge base.*

Top: *Used for just a few years and now in store are the Lockheed Hercules water-bombers. Since its first flight in August 1954, the Hercules is still being produced, having been upgraded over the years to the current C-130J model. Continuous production for fifty-seven years of the same aircraft is a record. Its main role is that of a medium transport to many of the world's air forces. However, a number of early-build aircraft have been declared surplus and have become water-bombers.* ***Lockheed C-130A Hercules N138FF*** *(c/n 3227), Tanker No 88 of International Air Response, is pictured at the company's base of Chandler Memorial, Arizona.*

Bottom: *International Air Response's* ***Lockheed C-130A Hercules N117TG*** *(c/n 3018), Tanker No 81, is pictured at Chandler having just arrived back with many of the company staff on board following a visit to a local air show.*

Top right: *Another company with stored C-130s is TBM Inc.* ***Lockheed C-130A Hercules N473TM*** *(c/n 3081), Tanker No 63, is pictured at Dinuba-Sequoia, California; it can now be found at Visalia, a few miles away.*

Middle and Bottom right: *Russia has vast areas of forests and the summers can get very hot, bringing a risk of fire. Since the forests are remote and lack road access, wheeled fire engines are of little use, so it is not a surprise that a specialist water-bomber has been produced. The airframe is based upon the final variant of the Antonov An-24 line, this being the An-32. This aircraft has nearly twice the power of the original and is fitted with a pair of Ivchenko AI-20M turboprops with a power output of 5,180shp. So big are these engines that they are now mounted above the high wing rather than under it, to cope with the propeller diameter. Pictured parked and also dumping a load of water is* ***Antonov An-32P UR-48086*** *(c/n 2901) at Zhukovsky, Moscow. The aircraft is owned and operated by the Ukraine-based manufacturer.*

63
N473TM

UR-48086

UKRAINE

21
21

27
27
N927AU

22
22
N922AU

Top left: *As well as operating a fleet of water-bombers, Aero Union of Chico, California, has for many years been in the forefront of the design and manufacture of the retardant tanks fitted to aircraft. The company's current fleet aircraft type is the Lockheed P-3 Orion, which it first introduced in 1990. It is the only commercial operator of the type in the world, as all the rest are military aircraft, used in their designed role as a maritime patrol. Pictured at base is* **Lockheed P-3A Orion N921AU** *(c/n 185-5098), Tanker No 21. The company has one of the smartest liveries of any operator in the world. It is contracted to the US Forestry Service and can operate in any of the US states except Hawaii during the long contract season. This will usually start in March in the south-east of the nation, moving over in May to the south-west, then to the west and north-west in July, and ending as late as December in southern California.*

Middle left: *Pictured at its Chico base is* **Lockheed P-3A Orion N927AU** *(c/n 185-5082), Tanker No 27 of Aero Union. When flight crews are on station they will work a six-day week and up to a maximum of 14 hours in any one day. Actual flying time is limited to 8 hours in one day and a maximum of 42 in a seven-day period. With some contracts being as long as 210 days, a relief crew will arrive to cover pilots going on holiday. As well as water-bomber operations the maintenance side of the company will service ex-US Navy P-3s from storage before they are delivered to new military customers, and will even train crews and ferry the aircraft to their new homes.*

Bottom left: *Pictured climbing out of Santa Barbara, California, is Aero Union's* **Lockheed P-3A Orion N922AU** *(c/n 185-5100), Tanker No 22. It has just been reloaded with retardant and is heading off to a fire. Note that on the belly tank the red colour of the retardant has marked the normally pristine white paint; the actual retardant is in the tank, which holds up to 3,000 US gallons. In the early days of aerial fire-fighting the chemical dropped was borate and for some time the aircraft were referred to as 'Borate Bombers'; however, this is no longer used, as it was highly toxic. Two brands are currently used. One is Fire-Trol LCG-R, which is basically a liquid ammonium polyphosphate with attaplugite clay added to the concentrate to suspend the colour and add visibility. It contains a corrosion inhibitor and iron oxide for colour, and is simply mixed with the required amount of water; 1 ton of liquid concentrate will produce 923 US gallons of retardant to be pumped directly into the aircraft. The other major brand used is Phos-Chek, which consists of 80% water, 14% fertilising salts and 6% colouring agent, corrosion inhibitors and flow conditioners. Despite the forests being covered in a red retardant, this will fade to a natural earth tan after a few days of direct sunshine.*

Below: *The strong red and white of Aero Union's Lockheed P-3 Orions is in marked contrast to the deep blue of the California sky at its Chico base.*

Top: *Before the Orion there was the Electra. This was a turboprop airliner built by Lockheed, which flew for the first time from the company's Burbank, California, factory in December 1957. Eastern Airlines was the launch customer. However, two problems occurred: one was the desire of the travelling public to fly in pure jets, and the other more serious one was a number of fatal accidents. The cause was eventually traced to excess vibration leading to wing root failure; the manufacturer designed a cure for this and all aircraft were modified. Many Electras are now in cargo service, as the last of the passenger examples have ceased operations. Air Spray of Red Deer, Alberta, has pioneered tanking the type for water-bombing, and pictured at base is* **Lockheed L-188A Electra C-FQYB** *(c/n 1063), Tanker No 488.*

Bottom: *The powerplants for the Electra are four Allison 501 turboprops with a power output of 3,750shp. Known in Air Spray service as the 'Longliner', it has the same capacity 3,000-US gallon tank as fitted to Aero Union's Orions; the Orion is of course a maritime patrol aircraft developed from the Electra. Pictured at its Red Deer base is* **Lockheed L-188A Electra C-FLJO** *(c/n 1103), Tanker No 482, owned and operated by Air Spray. This airframe began its life in 1959 and spent the following years with the US Federal Government, first with the FAA then with NASA, until it was stored in 1995 before joining Air Spray for conversion.*

Russian and Chinese-built aircraft

2

That great man Sir Winston Churchill once said that 'Russia is a riddle, wrapped in a mystery, inside an enigma'. He could have been talking about that nation's aviation industry prior to the August 1991 coup that led to the break-up of the USSR. Facts were few and hard to find, any success was trumpeted by the government-controlled media, and failures were hidden from view.

The aircraft the Russians produced were only operated by themselves, their satellite nations and by gifts to countries they wished to influence in the third world.

At one time all the civil aircraft in the USSR from international passenger traffic to domestic operations to crop-spraying were controlled by the vast organisation that was Aeroflot. Following the break-up most of the old republics have become independent states and took over the Aeroflot aircraft that were operated locally. Aeroflot itself is now a respected carrier operating some of the latest western-built Airbus and Boeing designs.

Russia has long had an aircraft industry, and while its aircraft may not have been as sophisticated as their western equivalents, they worked well in the harsh environment of bitterly cold winters, with some airports lacking the landing aids found in Europe and America.

One of the USSR's most important design companies is now in a different country, the newly independent Ukraine. Based in Kiev, the Antonov design bureau has produced the go-anywhere, carry-anything An-2 biplane. Its roles are whatever people have wanted it to do, and it has done them. The company also produced Russia's first-generation turboprop-powered airliner that replaced the post-war piston-powered ones. This range, starting with the An-24, has developed into military transports and survey versions, and ended with the An-32, having twice the power and thus giving it an excellent hot and high performance.

Antonov also designed and built a transport range starting with the An-8,

via the short-lived An-10 airliner, to the very successful An-12, a military and now commonly found civil lifter with the same layout as the Lockheed C-130.

The Ilyushin company produced the first Soviet post-war airliners in the form of the IL-12 and IL-14; both piston-powered, they have been replaced by the An-24 and its variants. The Ilyushin bureau also produced the IL-18, a four-turboprop-powered airliner that can still be found in limited service in parts of the world.

The Chinese aircraft industry has until recent times largely just made licensed – and, in the case of some military jets, unlicensed – copies of Russian aircraft. China has built the An-2, An-24 and An-12, and in each case has added refinements of its own; in the case of the An-24 it has taken it to a whole new development with the MA-60. As the years progress China's aircraft will begin to make more and more inroads into world markets.

Top: *Every nation needs a real workhorse aircraft that can go anywhere and carry anything that will fit through its door. In the west there are the Otter and the Beaver, and in the old Soviet bloc there is the Antonov An-2. This is a single-engined biplane powered by a 1,000hp nine-cylinder Shvetsov ASh-621R air-cooled radial piston driving a four-bladed propeller. Its first flight was in August 1947, and more than 3,400 were built by Antonov in the Ukraine before manufacturing moved to PZL in Poland, which produced another 11,500 airframes. It is used as a mini-airliner, a survey aircraft, a crop-sprayer, an air ambulance, a parachute drop plane, a glider tug and a fire-fighter. It can take off in as short a space as 500 to 600 feet and land in less than 600 feet. Pictured at the Chernoye Aircraft Repair Plant near Moscow is newly restored* ***Antonov An-2 RA-02766*** *(c/n 1G12544).*

Bottom left: *The repair plant at Chernoye takes semi-derelict airframes and does full restorations of the An-2. Pictured prior to delivery is* **Antonov An-2 RA-35541** *(c/n 1G11453) of Orenburg-based Oren Air. This carrier operates a mixed fleet up to the size of Boeing 737-800 series aircraft for its operations both domestic around Russia and into Western Europe. It can trace its roots back to 1932, but during the Soviet era all services for civil aviation came under the Aeroflot name. On the break-up of the USSR, after the failed August 1991 coup, the carrier resumed operations the following year under its present name. This An-2 is one of more than thirty PZL-built examples operated by the company, all fitted with twelve seats.*

Top: *This picture of five An-2 fuselages represents part of the very large stocks of airframes that the Chernoye Aircraft Repair Plant has stored for future orders and restorations.*

Bottom: *With the opening up of Russia as a capitalist society, the Chernoye plant has seen the need for a new aircraft, still piston-powered, at an affordable price, for the Canadian seasonal fly-in-fishing tours. It has produced an An-2 on floats for the home market, but has also recognised the potential for exports to Canada. Pictured at Zhukovsky, Moscow, is* **Antonov An-2V RA-17967** *(c/n 1G20951).*

Below: *The fitting of a turboprop to the An-2 has been a very slow development process. In more than twenty years only a very small number of aircraft have been fitted with a Glushenkov TVD-10B engine with a power output of 940shp. Pictured at Zhukovsky, Moscow, is* **Antonov An-3 629** *(c/n 1G18312-2629). Since the turboprop is much lighter than the big piston engine, the aircraft's centre of gravity has been changed, and to counteract this the fuselage has been extended by Antonov.*

Bottom: *The Chinese were the first nation to build the An-2 under licence. In the huge land mass of China, the need for a go-anywhere, carry-anything aircraft was realised as being very important. The Russians were at the time still friendly with China and in 1957 engineers arrived to help set up the production line. The first aircraft flew in December of that year, and was designated Y-5 (Yunshuji – meaning 'transport aircraft' – the fifth type). Pictured flying at Lanzhou is* ***Shijiazhuang Yunshuji Y-5 B-8238*** *(c/n 316409), operated by locally based China Northwest Airlines (now merged with China Eastern). Of note are the winglets on the top wing; these are known as 'tipsails' and are said to increase the climb rate, as well as affecting the aircraft's wake when crop-spraying.*

Top right: *When production of the Y-5 started in China it was at Nanchang and the engine was the Huosai HS-5, a licence-built version of the ASh-621R. It then moved to Harbin and, without any production there, moved on to Shijiazhuang, where more than 1,000 have been built. Pictured is a long line of Y-5s, built at both Nanchang and Shijiazhuang and operated by Xinjiang General Aviation at Shinezi in China's far west province.*

Middle: *One of the earliest original aircraft designed and built by the Chinese was the Harbin Y-11, a light transport with STOL characteristics. Its intended roles included transport, crop-spraying, survey and medical evacuations. The design and manufacture were at the Harbin aircraft factory in the far north of the country, and the first flight was in December 1975. Power for the aircraft was provided by a pair of Zhuzhou HS-6D air-cooled, nine-cylinder radial pistons with an output of 285hp each. These drove a variable-pitch propeller with two blades. Pictured at Shinezi is* ***Harbin Y-11 B-3870*** *(c/n 0502) operated by locally based Xinjiang General Aviation. This aircraft's normal task was crop-spraying and freight operations.*

Bottom: *July 1982 saw the first flight of a development of the Y-11, the Y-12. To give more power and performance a turboprop was fitted, the Pratt & Whitney PT-6A-27 with an output of 500shp. The fitting of a western-built powerplant would help export sales, as customers would know that spares would not be a problem. The aircraft has in fact been sold for both civil and military operations in some twenty-seven nations around the world, mainly developing ones. When used as a passenger airliner it has a capacity of seventeen seats, being both longer and wider than the Y-11. Pictured at Shinezi is a line of Harbin Y-12s operated by Xinjiang General Aviation, and used for photographic and survey work.*

Top: *Due to the government in Soviet Russia not understanding the words 'passenger choice', the Ilyushin IL-18 had a far longer front-line service life that its western counterparts, the Lockheed Electra and the Bristol Britannia. The same applied to operations in the Soviet bloc nations and developing countries under the communist influence. The last passenger IL-18s are now to be found only in ex-Soviet republics, and in ever-decreasing numbers. It has, however, found limited use as a cargo carrier for operations in developing countries. Pictured at Sharjah, UAE, is* ***Ilyushin IL-18GrM (SCD) 4R-EXD*** *(c/n 187009802), owned by Expo Aviation of Colombo, Sri Lanka, which operates both scheduled passenger services and cargo operations from both Colombo and Dubai.*

Middle: *The IL-18 first flew in July 1957 and was powered by four Ivchenko AI-20 turboprops. As well as civil operations, many could be found as test-beds for various systems, for electronic spying in an Aeroflot livery, and also as a maritime patrol aircraft under the designation IL-38. Pictured at Sharjah, UAE, is* ***Ilyushin IL-18D UN-75004*** *(c/n 186009202), operated by Irbis Air of Kazakhstan. It is alleged that this carrier was connected to the arms dealer Victor Bout and was used to ship illicit weapons to many of the world's trouble spots. Following the break-up of the Soviet Union the former republic was allocated the civil aircraft registration prefix 'UN'. This has since been changed to avoid confusion with operations by the United Nations, and this aircraft has now adopted the new national prefix of 'UP' and the new registration of UP-I1804 with its current owner Mega Airlines, based in the capital city Almaty.*

Bottom left: *At the time of the break-up of the Soviet Union it was thought by many that the Antonov An-8 was extinct, but they were proved wrong when the type began to appear in all sorts of places in the world, often carrying registrations of convenience and cargos that did not stand up to hard scrutiny. The role of the type was as a military freighter with large rear cargo doors that could be opened in flight to drop supplies for troops. First flown early in 1956, the power was supplied by a pair of Ivchenko AI-20 turboprops. It had a capacity for forty-eight troops or a 41,900lb (19,000kg) payload. Pictured at Sharjah, UAE, is* **Antonov An-8 3C-DDA** *(c/n OV.3420) in the livery of Mandala Air Cargo; the country of registration is Equatorial Guinea.*

Below: *Before the democratic election of President Ellen Johnson Sirleaf, who is also the first woman to be an African head of state, the nation of Liberia was renowned for being a haven of airlines and operators using it solely as a register of convenience; in many cases the aircraft would never visit the nation. Pictured at Sharjah, UAE, is* **Antonov An-8 EL-AKY** *(c/n OG.3410) in the colours of Santa Cruz Imperial Airlines, which was in fact based at Sharjah. Its fleet was a mix of Soviet-era cargo aircraft, the An-8 being the oldest. The Antonov OKB (Design Bureaux) withdrew the type's Certificate of Airworthiness and also design support in 2004, but various airframes could still be pictured at third-world nations following what should have been the end of legal flying operations.*

Bottom: *The logical development from the twin-engined An-8 was a larger four-engined version. This was the An-10, a passenger aircraft for Aeroflot that flew for the first time in March 1957. In its final form with a high-density interior it could seat 110 passengers, but following an accident in 1972 the type was retired from service and is no longer found in operation. The second development was the far more successful An-12 military freighter. This first took to the air in December 1957 and was powered by the same turboprop engines as the An-8. Its role and layout are similar to the American C-130 Hercules and it serves in the air forces of the former Soviet bloc as well as many non-aligned ones. With the demise of the USSR the aircraft have found their way on to the civil cargo market, mostly in third-world countries. Pictured landing at Dubai, UAE, is* **Antonov An-12BP S9-SAM** *(c/n 3341408) of Sharjah-based British Gulf International. The 'S9' registration prefix is from Sao Tome & Principe, a former Portuguese colony off the coast of west Africa and renowned as being a 'register of convenience' nation.*

Top: *Pictured at Sharjah, UAE, undergoing maintenance is* **Antonov An-12BP ST-AQQ** *(c/n 9346504) of Sudan States Aviation. This Khartoum-based company operates cargo services around Africa and the Middle East and appears on the European Union blacklist of carriers banned from European airspace and operations. As well as cargo operations both civil and military, the An-12 has been used as a test-bed for many applications including a Russian Air Force one that has an ejector seat test unit fitted aft of the tail fin; it can even be rotated to fire the seat out in different directions.*

Bottom: *The Republic of Armenia is one of the nations that emerged from the break-up of the Soviet Union. Based in the capital of Yerevan is Air Armenia, set up in 2003 and operating both scheduled and charter cargo operations. The scheduled one is from its base to Frankfurt-Hahn in Germany, and the charter services fly to all points in Europe, the Middle East, Africa and Asia as required. The An-12 can lift a payload of 20 tons and, with its military pedigree, has a rugged design allowing it to land on unprepared strips. This carrier advertises that it can transport two Jeep-type cars with ease within the cargo hold of the aircraft.* **Antonov An-12BK EK-11001** *(c/n 8346107) is seen with titles at Sharjah, UAE.*

Top right: *During the days of the USSR all civil aviation from international passenger services to cargo to crop-spraying came under the all-encompassing umbrella of Aeroflot. Following the USSR's demise many regions of the vast Russian republic started airline operations using former Aeroflot staff and equipment. Donavia was one, based in the city of Rostov-on-Don. In 2000 the company was purchased by Aeroflot and operated as Aeroflot-Don, but the old name of Donavia reappeared in 2009. Pictured at Sharjah, UAE, is* **Antonov An-12BP RA-12974** *(c/n 9346506) of Donavia Cargo. This airframe was originally operated by Aeroflot and the livery it now wears is still the basic Aeroflot scheme but with new titles.*

Middle right: *Based in Luanda, Angola, Savanair operated a pair of An-12s for ad hoc cargo services. Pictured at Sharjah, UAE, is* **Antonov An-12BP D2-FBY** *(c/n 8345510). The An-12 has one special feature missing on western-designed cargo aircraft, and that is the ability to carry a rear gunner. The turret, which could house a pair of 23mm cannons, was usually present on military versions and faired over on civil ones; however, it was not unusual to see a civil-operated aircraft with the turret still in place, albeit without the guns fitted. The An-12 has been widely exported and can be found flying in either civil or military operations in more than forty nations around the world.*

Below: *China has produced its own version of the An-12, known as the Y-8, and had planned to start production in the early 1960s, but with the rift in Sino-Soviet relations and the Cultural Revolution in China the work did not start until the following decade. December 1974 saw the first example in the air. Its powerplants were four Zhuzhou Engine Factory WJ-6s, these being reverse-engineered Russian AI-20K turboprops with an output of 4,250shp. It is possible to identify the Y-8 from the An-12 by the slightly longer nose, which is the same one the Chinese fitted to their licence-built Tu-16 bombers known as the Xian H-6. Pictured landing at Guangzhou in southern China is* ***Shaanxi Yun Y-8F-100 B-3102*** *(c/n 1002) operated by China Postal Airlines based at Tianjin.*

Top: *First flown from Kiev in December 1959, the Antonov An-24 was the first of what would be a long family of aircraft from either the parent company or from licensed developments in China. The An-24 was a high-wing airliner powered by a pair of turboprops and in the same class as the west's Avro 748/Fokker F.27/Handley Page Dart Herald and the Japanese YS-11. It entered passenger service in Aeroflot's Ukraine Division in October 1962 on the route from Kiev to Kherson. Still in service in the now independent Ukraine is* **Antonov An-24RV UR-BXC** *(c/n 37308902) operated by Zaporozhye-based Motor Sich Airlines. This aircraft is equipped with forty-eight passenger seats and is pictured at Moscow's Vnukovo Airport. The aircraft are owned by the Motor Sich Engine Manufacturing Company, which produces the giant powerplants for the Antonov An-124 and An-225 freighters.*

Middle: *China needed a modern aircraft to replace fleets of piston-powered Lisunov Li-2s (a Russian copy of the DC-3) and the Ilyushin IL-12 and IL-14s. Its choice was the Antonov An-24, to be built under licence. The first Chinese airframe flew in December 1970, but it took a very long time to get it into production due to lack of engineers, many having been purged during the disastrous Cultural Revolution. The licence-produced engine was underpowered, and twice the aircraft, known as the Y-7, failed its certification process. It was not until 1982 that it was awarded its type certificate. Pictured at Xian is* **Xian Y-7-100C B-3708** *(c/n 11705), operated by locally based Chang An Airlines. The '100' series has small winglets fitted and was designed in cooperation with a Hong Kong-based company to comply with British airworthiness requirements. The powerplants for this aircraft were a pair of WJ-5A turboprops with an output of 2,900shp each.*

Bottom left: *The '200' series of the Y-7 followed the opening of China to the west for business. This gave the Chinese aircraft manufacturers the ability to add western-built components to their airframes. The biggest change was in the engines – out went the locally built WJ-5s and in came a pair of Pratt & Whitney Canada PW-127C turboprops. Also new were the propellers, auxiliary power unit and avionics on the flight deck. The fuselage had a small stretch, and this meant that the seating capacity could rise to sixty passengers. Pictured at base is* ***Xian Y-7-200A B-3720*** *(c/n 0001), the first production aircraft operated by Chang An Airlines.*

Above: *The latest version of the Y-7 is now renamed the MA-60, 'MA' for 'Modern Ark' and '60' indicating the maximum seating capacity. It first flew in March 2000 and the type was certified by June of the same year. Once again the aircraft has a lot of western-built equipment. It has the same basic powerplant as the '200' series but with an increased fuel capacity. As well as this it features a two-man 'glass' cockpit with electronic flight instruments as well as weather radar and autopilot. Pictured at Johannesburg is* ***AVIC-1 MA-60 Z-WPK*** *(c/n 0303) of Air Zimbabwe. As well as with this country, the type is in service with airlines and air forces in more than a dozen countries around the world. The change from Xian to AVIC in the name relates to the political change of placing the Aviation Industries of China (AVIC) in control of the national airframe, engine and component manufactures.*

3 South American operations

Not too many years ago South America hosted much of the world's population of piston-powered transports. Sadly the decline has been quite rapid, and what is left is concentrated mostly in Colombia and, to a lesser extent, Bolivia. Many of the other nations in that vast continent are totally devoid of old gasoline-guzzling aircraft. In many cases they have been replaced by the early models of jet airliners such as the Boeing 727, 737 and the Douglas DC-9.

The building of new roads has all but killed off one of the classic aircraft roles in Bolivia, that of the meat freighter. The largest city, La Paz, is situated on the Altiplano or high plateau, and the airport is aptly named El Alto ('the high one') as it is 13,500 feet above sea level and the air is very thin, leaving any exercise out of the question for the visitor as a few days are needed to begin to acclimatise. With people living at altitude and the farms in the lowlands, a whole industry grew up to ferry freshly slaughtered cattle to the city. The aircraft used were often the oldest of the old, and the companies that operated them did not bother with maintenance as most did not last in business beyond a few years. These aircraft would come to grief due to a number of factors: the lack of maintenance, the workload of the pilots, their familiarity with the types flown, the lack of facilities on the farm strips they flew to, and the problems of flying in mountains without up-to-date navigation aids all played their part in the demise of some of the carriers and their aircraft. It was a new road that was the final nail in the coffin, and the sight of a C-46 powering out of La Paz to collect meat is now all but a memory.

Over in Colombia the one aircraft that can still be found in reasonable numbers is the classic Douglas DC-3. Its role is to service many small towns with both a freight and passenger service, often mixed together at the same time on the aircraft. The centre of operations is at Villavicencio, the last location before the start of the jungle.

Colombian aviation was also known for drug smuggling into the USA, but while this no doubt still goes on by other means, the virtual war footing of the US border agencies and their radars in fixed balloons and airborne radar-equipped aircraft has made the low-level flight entry across the water all but impossible.

As with the other remote places in the far north of the US and Canada, it is hoped that the roar of a round engine driving a propeller will be heard for some time yet in South America.

Below: *The Bolivian propliner scene has shrunk to almost nothing in the last couple of decades. Once the main industry for them was the transport of freshly slaughtered cattle from lowland farms up to the high-altitude city of La Paz, but new roads have seen this business wither away. Pictured at El Alto Airport, La Paz, is* ***Douglas C-47 CP-607*** *(c/n 12570) operated by Frigorifico Santa Rita. This airport, as its name suggests, is the 'high one', being 13,500 feet above sea level and set upon the Altiplano or high plateau. The main runway is 13,123 feet long, as the very thin air makes take-off more difficult. This C-47 was built in 1944, serving first with the US military; it was then sold to a Bolivian operator in 1945 and has operated there ever since with a number of different owners, operators and registrations.*

Below: *Colombia has South America's most thriving piston-powered scene. The main airfield for this is Villavicencio, which is the last town before the jungle starts and, with limited road access, is host to many operators flying into what are just unprepared strips; in fact, in one small town the aircraft land in the main street. These two views show* **Douglas C-47 HK-122** *(c/n 4414) on the ramp at Villavicencio when operated by locally based cargo carrier Lineas Aereas El Dorado. Built for the US military in 1942, this aircraft went to Colombia in 1947 and is still based at this location, now owned by Aliansa – Aerolineas Andinas SA.*

Top: *Another of the carriers based at Villavicencio is Air Colombia. Formed in 1980, and moving from Bogota in 1996, it operates three DC-3s, all in a utility configuration, which translates as both passenger and freight mixed. Flying in this part of the world is 'no frills' in the extreme, as for some flights passengers may sit upon the freight. Pictured in plan view from the airport control tower is* ***Douglas C-47 HK-3293*** *(c/n 9186). Of note is that both the airline name and the aircraft registration are painted on the top wing, a practice not to be found on airlines in Europe. This aircraft was built in 1943 and joined the Royal Air Force before moving to BOAC for its wartime operations. It went back into uniform with the RCAF from 1951 to 1976 and journeyed to South America in 1989.*

Bottom: *Colombian airline ADES – Aerolineas del Este – operated a single DC-3 in freight mode together with a number of light aircraft. Pictured on the ramp at its Villavicencio base is* ***Douglas C-47 HK-1149*** *(c/n 26593). A 1944-built airframe, it joined the US Navy and in 1959 the US Federal-operated CAA. At the end of 1965 it was passed on to the Colombian equivalent and in 1988 was sold on to the general market.*

SADELCA COLOMBIA
HK-3286

TAERCO COLOMBIA
HK-1315

TRANSPORTE AEREO MILITAR
2010

Top left: *Over the years Sadelca (Servicio Aerea del Caqueta) has operated at least seventeen different DC-3s from its Villavicencio base. Pictured there is* **Douglas C-47 HK-3286** *(c/n 6144). Built in 1942, it served first with the US military in Italy, before a post-war career with Pan American. It later moved to Panama before arriving in Colombia.*

Middle left: *Taerco (Taxi Aero Colombiano) operated a single DC-3 for a few years. Its main operations revolve around a pair of single-motor Cessna aircraft. Pictured at its Villavicencio base is* **Douglas C-47 HK-1315** *(c/n 4307). This airframe was built in 1942 and spent most of the wartime period with the US military based within the continental USA. In 1946 it was sold to the American civil market, then moved two years later to Brazil, staying until 1966 when it moved to Colombia.*

Bottom left: *The Fuerza Aerea del Paraguay (Paraguayan Air Force) has flown the DC-3 since 1954 and still does to this day. More than thirty examples have been operated, with deliveries from ex-USAF stocks over a number of years. Pictured on the ramp at the capital Asuncion is* **Douglas C-47 2010** *(c/n 32630). This aircraft, built in 1945, has spent its entire life in uniform, first in the USA, then Paraguay.*

Below: *The old-established German manufacturer, Dornier, has since the end of the Second World War built a number of rugged STOL transports. First flown in February 1966, the Do 28D Skyservant was a completely new design from the previous Do 28, that having been derived from the single-engine Do 27. It was powered by a pair of Lycoming piston engines with an output of 380hp each. Almost all the production was delivered to air forces or governments around the world. One of the very few civil operators of the type is Villavicencio-based Trans Oriente (Transporte Aereo Regular Secundario Oriental SA), which flies a mix of passengers and cargo, the aircraft having a capacity of ten seats. Pictured at base is* **Dornier Do 28D-2 Skyservant HK-3991** *(c/n 4148).*

Bottom: *In Bolivia a lowland farmer would contract for a freighter to take meat to La Paz. The aircraft would take off and radio that it was en route, and the cattle would be taken from the fields, killed and cut up. The aircraft would arrive at the farmer's strip, usually unprepared and with no navigation aids, so all flights would be under VFR (Visual Flight Rules) conditions. The carcasses would be loaded on to the aircraft and flown to La Paz, unloaded onto trucks and taken to the market or even sold to locals at the airport. The whole process would be run as quickly as possible as neither the farm, the aircraft nor the delivery trucks used any form of refrigeration. With the sharp decline in this type of operation many of the former meat haulers sit awaiting a new role. On such is* **Douglas DC-6A CP-1282** *(c/n 45530) of La Cumbre (Transportes Aereos La Cumbre), as it sits at La Paz getting dustier as the years go by.*

Top: *The DC-6 has proved to be the big piston with the longest life, and still has a role in cargo operations in various parts of the world as well as its water-bombing tasks. Pictured at its Villavicencio base is* **Douglas DC-6A HK-4046X** *(c/n 43708) of general cargo carrier Air Colombia, which has since sold it on to a Venezuelan company.*

Bottom: *Pictured at its La Paz, Bolivia, base is* **Convair 340 CP-2026** *(c/n 249), then operated by CAT (Carga Aero Transportada). This airframe started its life as a C-131 with the USAF in 1955 and served until 1977, when it was put in store at Davis-Monthan AFB in Arizona. It moved to South America and now operates for a private owner. The colour scheme, or rather lack of it, is a common feature in South America; as aircraft pass from one owner to another they do not have to invest in an expensive repaint. Still visible on this aircraft are the faded US Air Force letters on the nose, and it is fitted with an air-stair on the entry door; it also has a rear freight door partly open.*

Top: *The Curtiss Commando is perhaps the aircraft most associated with operations in South America, where it has achieved a reputation as the 'tramp steamer of the skies'. Pictured at Cochabamba is* **Curtiss C-46 Commando CP-1655** *(c/n 33294), then operated by SAC (Servicios Aereos Cochabamba). Since then it has had several owners and one crash In 1999, when on a cargo-hauling job from La Paz, it lost an engine and made a forced landing in a swamp; the C-46 was well known for having little or no single-engine performance. For three years it lay there before a rescue effort saw the airframe dragged from the swamp and temporary repairs made to ferry the aircraft to nearby San Borja for more permanent work to be carried out. In 2004 it was flown, with the undercarriage down, back to La Paz and grounded, as both propellers and powerplants were time-expired. The aircraft is currently for sale.*

Bottom: **Curtiss C-46 Commando CP-1319** *(c/n 22428) is pictured at La Paz. From this view it looks as though the aircraft has reached the end of its active life and is being parted out for spares. However, being Bolivia, it was merely being stored, albeit for some years, before it once again took to the air for general cargo hauling.*

Top: *When an aircraft is being rebuilt to fly in Europe or the USA the one thing that you expect is that it would take place in a well-equipped hangar. This picture of* ***Curtiss C-46A Commando CP-1080*** *(c/n 26771) at La Paz shows how a rebuild takes place Bolivian-style, out in the open with wooden frames holding up the aircraft. This restoration was eventually completed and the aircraft flew assorted cargo services around Bolivia. It is currently awaiting a new lease of life, not having flown for some time.*

Bottom: *The ramp at La Paz shows a parked Douglas DC-3 and a Curtiss C-46 awaiting attention. Of note is the surface of the ground; despite being a major airport the surface is rough and broken with lots of small stones.*

Top: *In terms of numbers sold, the Fokker Friendship has been the most successful of the 'DC-3 Replacements'. As well as the Dutch production, from the earliest days it was licence-built in America by Fairchild at its Hagerstown, Maryland, site. That company's first example flew in April 1958, and it later stretched the fuselage to produce the FH-227, a variant able to seat fifty-two passengers. The powerplants on all versions, both Dutch and American, were Rolls-Royce Dart turboprops. Uruguay has no large piston-powered aircraft in operation, but propellers can be found turning on turboprops. Pictured in the capital, Montevideo, is* **Fairchild F-27F Friendship CX-BRT** *(c/n 99), leased to Air Atlantic Uruguay. This 1963-built aircraft started its life as an executive transport for one of the Rockefeller family, later serving the same role with several American corporations.*

Bottom: *As well as civilian roles, the Friendship also has a number of military operators. One such is the Peruvian Coast Guard, and seen here at its Lima base is* **Fokker F.27-600 AB584** *(c/n 10322). This airframe was not built as a maritime patrol aircraft but as a passenger-configured airliner for Ansett of Australia in 1967. Since then it has operated in Spain and Canada before joining its current owner in 1995. When this picture was taken it was waiting for an engine to be returned from being overhauled.*

Top: *The ultimate variant of the Antonov twin-turboprop range has proved to be a very popular aircraft in South America. With its STOL (Short Take-Off and Landing) performance due to the massive power of a pair of 5,180shp Ivchenko AI-20DM turboprops, it has the hot and high performance needed for regions of the continent. Pictured at the Colombian capital, Bogota, is* **Antonov An-32B HK-4011X** *(c/n 3208), operated by SAEP (Servicios Aereos Especializados en Transportes Petroleros). As the company name suggests, its role is in support of the nation's oil industry. Of note is the company logo on the tail – the 'A' in 'SAEP' is in the form of an oil rig.*

Bottom: *Flying a mix of passengers and general cargo from Villavicencio, Colombia, is Sadelca (Sociedad Aerea del Caqueta)* **Antonov An-32B HK-4006X** *(c/n 3049). As well as the Russian-built turboprops, the carrier also still operates four of the evergreen Douglas DC-3s. One of the abilities of the An-32, with its excess of power, is to lift a 6-ton load from a rough jungle strip. This has led to some operators of the type allegedly being implicated in exporting what Colombia is best known for, illicit drugs in the form of pure cocaine.*

Top right: *Peru has long been an operator of Russian-built equipment for its armed forces, as well as products from the USA and Europe. The An-32 is in service with four different branches of the military, all seen here at their Lima-Callao Airport bases. First is* **Antonov An-32B EP833** *(c/n 3104), operated by the Aviacion de Ejercito Peruano (Peruvian Army), with a role of logistical support.*

Middle right: **Antonov An-32B AT-531** *(c/n 3408) is operated by Escuadron Aeronaval 32, part of Grupo Aeronaval No 3 of the Servicio Aeronaval de la Marina Peruana (Peruvian Navy). As with the Army example, its role is that of transport.*

Bottom right: *Over the years the Fuerza Aerea del Peru (Peruvian Air Force) has operated a large number of the type.* **Antonov An-32B FAP-323** *(c/n 2809) is flown by Escuadron de Transport 842, part of Grupo Aereo de Transport No 8 with its headquarters at the airport. It is of note that some of the air force aircraft are equipped with seats for passenger flights while others are just for cargo.*

833
EJERCITO DEL PERU

NAVAL

FUERZA AEREA DEL PERU
323

Top: *Finally, the Policia Nacional Peruana (Peruvian National Police) has operated an air wing since 1983 and many of its fixed-wing assets are aircraft that have been seized from drug-runners and impounded. While they do not usually have the relevant maintenance records for these aircraft, they can be flown for a period of time then just cast aside on the basis that they cost nothing to acquire and the only costs have been fuel and a repaint into police livery.* ***Antonov An-32 PNP-234*** *(c/n 2502) is a legitimately purchased airframe and shows off the attractive livery of the Police Air Wing. It is operated by Escuadron 500 at Lima Airport.*

Bottom: *As well as the military operations of the twin Antonov aircraft, the airlines of Peru have operated almost all of the versions built by the Kiev-based company. Pictured with its two Ivchenko AI-20 turboprops turning as it taxies out to depart from its Lima base is* ***Antonov AN24RV OB-1650*** *(c/n 37308802) of Aero Condor. This carrier operates eight different aircraft types within a fleet of just thirteen airframes, ranging from five to fifty-seat capacity. This Antonov is equipped with forty-eight passenger seats and was built in 1973. The company commenced services in 1975 and at one time flew scheduled domestic services, but today just flies charters, sightseeing and medical evacuation services.*

Below: *The main difference between the An-24 and An-26 is that the latter has a dual-hinged rear cargo ramp door whereby small vehicles can be driven straight into the fuselage; it also has two ventral fins on each side of the loading ramp. Pictured at its base of Lima, Peru, is* **Antonov An-26B OB-1778P** *(c/n 14205) of ATSA (Aero Transport SA). This aircraft is the only freighter operated by the company; however, the interior can be fitted for up to forty-one passengers and their luggage or just nine passengers and 2,600kg of cargo. The 'B' version of the An-26 is fitted with a rollamat floor to ease the handling of cargo pallets. Founded in 1980, the company operates a fleet of modern turboprops for charters, VIP services and medical evacuations.*

4 Cargo services

The world of air cargo has at one extreme the vast fleets of wide-body jets operating next-day deliveries of small parcels to the USA and world-wide, and at the other an old C-46 Commando hauling burgers, beans and bed linen to native American reservations, with everything in between.

It was once the fate of retired airliners to be converted to carry freight; now almost every new airliner is offered with a cargo version, or has the capability for cargo pallets in its belly hold.

This book is of course concerned with the old, not the new, and among the oldest are the Second World War-vintage Curtiss C-46s, now only to be found in a few places in Alaska, Canada and Bolivia. The aircraft can uplift more cargo than the Douglas DC-3/C-47, but due to their low numbers there are no plans to re-engine them with turboprop power, a process that has given the DC-3 a whole new lease of life.

The series of Douglas-built four-engine aircraft, the DC-4/6/7, still hang on in lower numbers each year, with again no prospect of turbine conversion.

There are two main problems that will sadly ground the piston-powered transports. One is the eventual shortage of engine spares; these engines have not been made for years and a unit can only take so many overhauls before a new one is required. The other problem, perhaps more serious, is the lack of AVGAS (aviation gasoline) in many parts of the world, together with its high price when it is available, making the cost of operating the aircraft too high for the market to stand.

However, one advantage that the older converted airliner has, be it piston or turbine, is that the airframe is cheaper to acquire and can sit unemployed for longer periods than a newer, more expensive jet.

When the call comes for an urgent cargo delivery, they are ready. The load they carry can be anything that is needed quickly, often jobs for the motor industry, where vital parts to keep the production line running are needed if the normal just-in-time process has failed. It is cheaper for a supplier to stand the cost of short-notice air cargo than allow a production line to stop even for a few hours.

Where there is a need for a product quickly, there will always be a company ready to fly it to its destination.

Top: *The history of the Guppy outsize transports runs in parallel with that of the American space programme. NASA found that most of the very large bulky components were manufactured in California, home of most of the aerospace industry, but would be launched from Florida. The components were far too big to travel by road or rail, and the cost and time of the journey by barge via the Panama Canal was not acceptable. NASA was presented with the idea of using surplus Boeing Stratocruiser airliners and its military variant, the C-97, with a rebuilt outsize fuselage. The process of loading was made easy by the entire nose section being hinged and able to be swung open. This project was given the go-ahead and proved to be the answer to NASA's problems. The Guppy programme grew as companies such as Airbus found that it was the perfect way to move all its components from factories across Europe to the assembly plant at Toulouse in the south of France. NASA still uses a Guppy, albeit a newer turbine-powered one, for the same task of moving outsize components from where they are built to where they are needed. Pictured at Biggs Army Air Field, Texas, is* ***Airbus Industries 377SGT-201F Super Guppy N941NA*** *(c/n 004) with the NASA logo on the fin. The original manufacturer of the type, Aero Spacelines, sold the design and manufacturing rights to Airbus.*

Bottom: *With the success of the Beaver and Otter, de Havilland Canada saw a need for a large twin-engined transport with the same STOL characteristics that had marked out its previous designs. The new aircraft was called the Caribou and flew for the first time from Downsview in July 1958. The powerplants were a pair of Pratt & Whitney R-2000 Twin Wasp air-cooled radial pistons with an output of 1,450hp each. The design had an upswept tail, the fuselage floor was at truck height for ease of loading, and the rear door was able to be opened in flight for supply-dropping or parachuting. Almost all of the 300-plus airframes built went to air forces around the world with just a small number being operated by civilian companies. As they have been declared surplus from military service a few can be found hauling freight. Pictured at Madrid-Cuatro Vientos is* ***de Havilland Canada DHC-4 Caribou EC-GQL*** *(c/n 258) operated by locally based Avinsa Lineas Aereas.*

Top: *A cargo aircraft that grew from a glider was the C-123. Its origin was to found in the Chase XG-20 cargo glider. Following the demise of that company the design and manufacture were taken over by Fairchild. The first flight was in September 1954, powered by a pair of P&W R-2800 radials outputting 2,300hp each. It was bought by the USAF and also served with the air forces of Saudi Arabia and Venezuela; later, surplus USAF aircraft found their way to other air forces, including that of Thailand. A small number found their way onto the civil register, with one converted to a water-bomber. Pictured at Kingman, Arizona, is* **Fairchild C-123K N546S** *(c/n 20064). This aircraft has the extra General Electric J-85 jets under each wing, rated at 2,850lb of thrust to boost performance. The aircraft was undergoing trails of guiding supply drops with a GPS (Global Positioning System) fitted for accurate delivery.*

Bottom: *With its box-like fuselage, the Short Skyvan is an ideal aircraft for moving bulky items, with a payload of up to 3,500lb. The aircraft's rear opening cargo door is 6ft 2in wide, 6ft 6in high and 19 feet long. Pictured at Anchorage, Alaska, is* **Short SC-7 Skyvan 3 N731E** *(c/n SH-1853), owned by North Star Air Cargo based in Milwaukee, Wisconsin, but on lease to Alaska Air Taxi. It is one of four owned by North Star, one of which is used by a sky-diving school. The early models were underpowered but the current fitment of Garrett AiResearch TP-331 turboprops with an output of 715shp each has cured this. However, the aircraft does have quite a short maximum range of just 400 miles (710km).*

Below: *UK-based Atlantic Airlines is the largest user of the Electra in Europe, and operates scheduled cargo services, usually at night, as well as ad hoc ones. Typical of the latter would be just-in-time parts for motor industry production lines, while a scheduled flight will collect and deliver cargo at several different locations during the course of the service. Pictured at Coventry is* **Lockheed L-188C (F) Electra G-LOFE** *(c/n 1144), taking off with the undercarriage still down. The second picture of the same aircraft has it in an entirely different role from that of a night freighter. The UK Maritime & Coastguard Agency has a contract with the airline to provide an Electra fitted with spray equipment to fly over oil spills at sea and, with the aid of chemical dispersants, break down the oil to stop it arriving on the shoreline. Note in this photograph the temporary spray bars fitted at the rear of the fuselage as it sprays water in a demonstration of this valuable role.*

Top: *With the cost and availability of AVGAS (aviation gasoline) becoming more and more an issue for the operators of piston-powered fleets, it is inevitable that aircraft would be re-engined or that turbine-powered aircraft would join them. So it is with Buffalo Airways of Yellowknife, NWT, which has added a pair of Electras to its fleet for freight operations. The type can lift 33,000lb of payload, and for the mining, oil and gas exploration companies that rely on Buffalo this is a big jump from the 20,000lb load of a DC-4, the next biggest freighter. With its four Allison 501 turboprops each rated at 3,750shp, a range of 2,200 miles and a cruising speed of 390mph, the number the company operates will no doubt grow. Pictured outside the company hangar is* ***Lockheed L-188C(F) Electra C-GLBA*** *(c/n 1145).*

Bottom: *The first of the three Douglas four-engine pistons still earns its keep in parts of the world, albeit in small numbers. Pictured at Brisbane's Archerfield is* ***Douglas DC-4/C-54 Skymaster VH-PAF*** *(c/n 27352) when operated by locally based Pacific Air Freighters. This was the only one of its type in Australia and its role was ad hoc charters, such as freight for oil companies to Papua New Guinea and food to the Solomon Islands. The aircraft was built in 1945 and served with the US military until 1972. In the early years of her subsequent US civil career there were allegations of smuggling and the aircraft was impounded for periods of time. The DC-4 arrived in Australia in 1995 and in recent months has flown to Albion Park to join the fleet of the Historic Aircraft Restoration Society, which also operates Australia's sole Super Constellation.*

Top: *Canada's largest user of the DC-4 is Buffalo Airways. This company has achieved world-wide fame with the television series Ice Pilots, which shows its operations in both bleakest midwinter, when daylight hours are few and the temperature is well below freezing, and summer days of almost perpetual daylight. The DC-4 can haul 20,000lb of payload or be configured as a water-bomber with an under-fuselage tank for the retardant. One is kept and equipped with spray bars to eradicate insects and bugs with pesticides and biological agents or antipollution products to clean environment-harming spills. Pictured at base is* **Douglas DC-4/C-54 Skymaster C-FIQM** *(c/n 36088), No 57. This aircraft has recently been used at Mackenzie, British Colombia, with an under-fuselage mechanism to drop, for a TV programme, a reconstruction of the type of bouncing bomb used by the Lancasters of 617 Squadron in their famous 1943 raid on the German dams of the Ruhr.*
The second photograph, taken at Buffalo's second base at Hay River, shows **Douglas DC-4/C-54 C-GBNV** *(c/n 35988), No 56. This aircraft is missing one of its propellers and is configured at the moment as a freighter, but with less than a day's work it could be fitted out as a water-bomber.*
The third picture is **Douglas DC-4/C-54 C-GCTF** *(c/n 27281), No 58, seen at Yellowknife; note that all three have different variations of the livery.*

Top: *Last of the Buffalo group of pictures is a shot from Hay River showing three of the DC-4 noses. The two red ones are in the livery of their previous owner, Aero Union of Chico, California.*

Bottom and Top right: *Founded in 1986, Brooks Fuel of Fairbanks, Alaska, operates DC-4s as flying fuel tankers to take petrol and diesel to remote settlements, mines and oil sites that have no road access. Without an aerial supply these locations would not be able to heat their sites, run their generators or drive their vehicles. The DC-4 can deliver 2,860 US gallons. Pictured at base is* **Douglas DC-4/C-54 N3054V** *(c/n 10547) with all four pistons pounding as they warm up before the start of another fuel delivery flight. The second shot is the aircraft taxiing out to the main runway at Fairbanks. Its livery is basically that of the previous owner, Aero Flite. Built at the start of 1945, this aircraft has served with the US Army Air Force, the Royal Air Force and the US Navy, and was once also a water-bomber with Aero Union.*

Middle right: *What may look to some people like a junk yard is in fact the storage area of Brooks Fuel's ramp. It shows two of the company's DC-4s in store, together with its* **DC-7C N90251** *(c/n 45367). This, the largest of Brooks' fleet, has not flown since 1999 following an engine fire but, as with any old aircraft in Alaska, never say that it will not be brought back into service if required. As well as the ramp at Fairbanks, the company has a further four examples of the Skymaster in store at Falcon Field, Mesa, Arizona, awaiting the call to fly to the cold north-west in America's last frontier.*

Bottom right: *All three of the Convair twins still can be found earning revenue to this day. The very first was a single model 110, a tricycle-undercarriage, thirty-seat airliner first flown in 1946. Airlines wanted a bigger aircraft, so came the model 240 (with two engines and forty passengers); piston-powered, it was one of the first small airliners to be pressurised. The next model was the 340 (the original name base being no longer in use as it was not three-engined). This had a 54-inch stretch of the fuselage and a 14-foot wingspan increase, and was again piston-powered but with a higher output on the engines. The last of the piston models was the 440, known as the Metropolitan; it was the same size as the 340 but had greater weight capability and optional weather radar in the nose, making it slightly longer. Pictured at Red Deer, Alberta, is* **Convair 240 C-GTFC** *(c/n 279) in the livery of carrier Transfair, from Sept-Iles, Quebec, but in fact owned by Buffalo Airways. That company had bought the aircraft for its two Pratt & Whitney Double Wasp R-2800 engines, so they could be used on its Canadair CL-214 fleet.*

DOUGLAS DC4

N51802

Trans Fair

Top: *Kelowna Flightcraft now has all the design rights for the Convair twins and currently fits turbine engines, converts to water-bomber status or stretches to make the CV 5800 cargo carrier. To do all this the company needs a supply of raw material in the form of airframes to convert. Pictured at its Kelowna, BC, base is* **Convair 440 N4753B** *(c/n 340). This cargo-converted aircraft was built in 1956 for the USAF as a C-131E and is one of a number of aircraft owned by the company for either spare-part reclamation or full conversion.*

Bottom: *Photographed at Anchorage Airport, Alaska, is* **Convair 240 N153PA** *(c/n 304) in the colours of locally based Desert Air. This aircraft can haul 7,500lb of payload, can land on strips just 3,800 feet long, and has a bulk capacity of 2,100 cubic feet. The company started operations during the 1990s in Utah flying components for the motor industry across the USA and Canada before setting up in Alaska in 2001. It has a fleet of just two aircraft, the other being the everlasting DC-3. Desert Air claims that it will fly cargo direct to more than 200 locations within the largest state of the union. This Convair was built as a T-29 trainer for the USAF in 1953 and served with the air force until 1975, when it went into store at Davis-Monthan AFB, Arizona. It took a civil identity four years later and has served several freight and passenger companies over the years.*

Top right: *Seen on the ramp at Cape Town is* **Convair 580(F) ZS-SKL** *(c/n 458) of Johannesburg-based Skyhaul. This company operates four 580s but, as can be seen, does not have its name on the aircraft. This airframe was built as a model 440 for Australian airline Ansett in 1957 and was converted to a turboprop 580 in 1967. The following years saw the plane operate for a number of companies in the USA and Spain before it arrived in South Africa.*

Middle right: *Only a small number of DC-6s are to be found flying in America's 'lower 48', the area of continental USA, excluding Alaska and Hawaii. Nord Aviation, based at Santa Teresa, New Mexico, operates a single example, as well as three DC-3s. Pictured at the company base is* **Douglas DC-6A N620NA** *(c/n 44677), built in 1954 and civil-operated all its life with many owners during that time. The current owner, who flies night cargo, has had it since 1993.*

Bottom right: *The area of Miami and the local airfields of south Florida used to be a hive of propliner activity, flying to the islands of the Caribbean and to South America. However, the American authorities banned some countries from entering US airspace due to safety concerns, and these included Haiti, the Dominican Republic and some of their neighbours. The vast majority of the area's piston-powered aircraft can now be found at Opa Locka, just*

to the north of Miami, having moved there when Miami Airport evicted all the old props that had made it such a haven for visiting enthusiasts. The most established of the operators at the new location is Florida Air Transport, which flies all of the Douglas four-engine designs, the DC-4/6/7s. Pictured at nearby Fort Lauderdale Executive Airport is ***Douglas DC-6A N70BF*** *(c/n 43720) with its cargo door open ready to take its next shipment.*

Below: *Once the biggest US operator of the DC-6, Anchorage-based Northern Air Cargo has only in the last few years retired its fleet. The company had bases at Fairbanks as well as Anchorage, with aircraft that hauled freight and a number specially adapted for the delivery of bulk fuel to remote settlements; these carried the name Northern Air Fuels. Photographed on the ramp at Anchorage, on one of the company's last commercial trips, is* **Douglas DC-6A N6174C (c/n 44075)**. *So good is the DC-6 that this aircraft is now operated by Fairbanks-based Everts Air Cargo. One of NAC's other aircraft,* **Douglas DC-6BF N867TA** *(c/n 45202), is seen at the Fairbanks depot with all four Pratt & Whitneys pounding and raising a dust storm behind it. This aircraft was one of two in the fleet that had been converted to a swing-tail version; the whole unit could open and was therefore able to load long items, for example a ship's propeller shaft, which could not fit through the normal cargo door.*

Top right: *The United Kingdom's sole DC-6 operator has for many years been Coventry-based Air Atlantique. The pair of aircraft have been used as ad hoc freighters, one-off jobs including both being chartered for use in the James Bond film Casino Royale.* **Douglas DC-6A/B G-SIXC** *(c/n 45550) is pictured on take-off at its base. This was one of the last of the line when it was completed in 1958, being first operated in Taiwan, then Laos, before heading to the USA and eventually the UK in 1987.*

Bottom right and top page 84: *Sister ship at Air Atlantique is* **Douglas DC-6A G-APSA** *(c/n 45497), pictured at base first cavorting for the camera at an air show with steep topside passes, then showing off one of its roles, that of dealing with oil spills off the coast by spraying dispersant on them. Note the spray bars at the rear of the fuselage during a demonstration when water was being used; this form of operation is taxing for the two pilots as they have to fly very low, as shown. This aircraft was built in 1958 for a Canadian airline; it joined a UK company the same year before service in Saudi Arabia, then Yemen. It joined its current owner in 1987.*

ATLANTIC
G-SIXC

Air Atlantique

Bottom: *One of Alaska's main industries is that of fishing, both deep-sea trawling and for the millions of salmon that make their final trip to spawn. There is just a short season for this amazing journey as the fish can swim hundreds of miles up rivers, against strong currents and jumping rapids to lay eggs or fertilise them depending upon their sex. It is believed that the fish's sense of smell brings it back to the place of its birth. With so many fish available to catch, an industry has grown up to fly the freshly caught fish to markets and canning factories. Photographed at Kenai, Alaska, is* **Douglas DC-6A N500UA** *(c/n 44597), operated by Universal Airlines. The company has a fleet of three DC-6s and is based at Victoria Regional Airport, Texas, but all the aviation activities keep the aircraft in Alaska year round. When the season is over the fleet will just park up at Kenai until the following year.*

Right: *Currently the largest operator of the DC-6 in the world, Everts of Fairbanks, Alaska, has a fleet of fourteen, operating between the two main units of the company, Everts Air Cargo and Everts Air Fuel. The cargo operations fly whatever will fit into the doors of the DC-6 up to a maximum weight of 30,000lb, varying with the distance to destination, the size of the runway upon arrival and whether fuel is need to be carried to operate both sectors of the flight. For people living in the remote parts of the state, the aircraft brings all their requirements from the lumber to build with to the food they eat, and everything in between. The company also has a base at Anchorage and operates scheduled cargo flights to eleven destinations including King Salmon, Nome and Bethel on a daily basis for six days a week – Sunday is the day off. Small towns are served only two or three times a week. One of the DC-6 fleet is based at Deadhorse in the far north of*

the state for operations to Barrow and other ad hoc locations during the bulk of the year. Pictured at the Anchorage depot is ***Douglas DC-6BF N151*** *(c/n 45496). This airframe is an ex-water-bomber purchased from Conair of Abbotsford. The second picture is a view of the company ramp with a DC-6 and one of its workhorse C-46s awaiting their next cargo load.*

The air fuel division of the company keeps both a C-46 and a DC-6 at Kenai Municipal Airport, from where they will ferry petrol, diesel and heating oil to settlements, mining camps and any location that has a runway long enough. The DC-6 can haul up to 5,000 US gallons but usually flies with less due to landing weights on some of the gravel strips that serve as runways. The crews here will fly both types, being dual-licensed. Pictured at the Kenai location is Everts Air Fuel ***Douglas DC-6B N444CE*** *(c/n 45478). This 1958-built aircraft first served in France as a passenger plane with UTA before moving to Belgium, then Canada, serving there as a water-bomber in 1975 before joining Everts twenty-two years later in 1997. It carries on the nose the name 'Spirit of America'.*

These two photographs show the noses of pairs of DC-6s, the first two from Northern Air Cargo in their final days with the type at Anchorage, and the second a duo from Universal Airlines at Kenai, awaiting the start of the next season hauling fresh salmon.

Top: *At first glance the C-46 looks to have the same proportions as the DC-3, but it is a much larger aircraft. It has a 108-foot (32.91-metre) wingspan, against the DC-3's 95 feet (28.96 metres), is 76ft 4in (23.26m) long as against 64ft 5in (19.63m), and 21ft 9in (6.62m) high, against the DC-3's 16ft 11in (5.16m). In capacity it has a gross weight of 40,00lb (18,144kg) compared to the DC-3's 28,000lb (11,431kg). The users will soon be able to be counted on the fingers of one hand, with northern Canada and Alaska being the last strongholds. In Canada First Nations Transportation is based at Gimli in Manitoba, north of Winnipeg, and operates a pair of Commandos to service the native population on their reservations; they will carry all the supplies needed by these communities from baked beans to bed sheets. Pictured at base is* ***Curtiss C-46 Commando C-GIBX*** *(c/n 22472) with just the minimum of titles upon the fin. This aircraft was built in 1944 for the US military and following the conflict's end joined the US civil register as a freighter; it moved north to Canada in 1985, where it has been registered ever since. During the mid-1990s it spent a two-year period in Africa operating relief flights delivering food aid.*

Bottom: *First Nations Transportation's other C-46 is still in the basic livery of its previous operator, Buffalo Airways, but no longer carries any titles.* ***Curtiss C-46 Commando C-GTPO*** *(c/n 22556) is pictured at base with its cargo door open and typical loads awaiting shipment on the ramp. Built in 1945 for the USAAF, it was declared surplus and sold in 1950, moving to Canada in 1986. In October 1989, while operating with Winnipeg-based Air Manitoba, it crashed on take-off at Pickle Lake, Ontario, on a fuel delivery flight. The port undercarriage collapsed, further damage was done to the wing and spar, and the port engine suffered a post-accident fire. Due to its location and the age of the aircraft, it was stripped of many parts and written off. However, such was the value and capability of the type that Buffalo Airways sent a team to restore the aircraft sufficiently to ferry it to Yellowknife, where a full restoration followed.*

TRUCK-AIR-EDM

-FAVO

C-GTXW

Left: *Buffalo Airways currently operates a pair of the big Curtiss cargo haulers. They can lift up to 14,000lb of mixed freight and deliver it to the many short, unprepared gravel strips in the Northwest Territories of Canada. Pictured with both its Pratt & Whitney R-2800 radials pounding,* **Curtiss C-46D Commando C-FAVO** *(c/n 33242) taxies out at its Yellowknife base for another cargo delivery. Note under the cockpit the logo of German flag-carrier Lufthansa, a throwback to the 1960s when this aircraft was owned by Capitol Airways and was leased by Lufthansa for cargo services, becoming a frequent visitor to London and other European cities. The aircraft has since then, via various owners and operators, kept the logo in place.*
The second picture of C-FAVO, again at Yellowknife, shows it being loaded with its cargo. Note that a heater has been connected to the starboard engine to warm the oil prior to starting it, which is normal practice for winter operations; although in fact this photograph was taken early in the month of May, some areas of snow are still visible.
The company's other **Curtiss C-46A Commando C-GTXW** *(c/n 30386) is pictured being pushed out of the maintenance hangar, with the tow-truck at the rear. Note the two people walking by the wing tips to ensure that they do not collide with anything else on the crowded ramp. This aircraft is in quite a different livery style from the other one: the wings are green, the Buffalo name on the cabin roof has a different layout, and the typeface used on the registration and its size are also changed.*

Top: *With a fleet of four C-46s, two with air cargo and two with air fuels, Fairbanks-based Everts now has the largest fleet of the type in service in the world. Pictured at base is* **Curtiss C-46F Commando N1837M** *(c/n 22388) in the livery of Everts Air Fuels. This aircraft is named 'Hot Stuff' and is being towed along the ramp to the hangar to have the port Pratt & Whitney R-2800 reconnected. The fuselage is fitted with a tank capacity of 2,100 US gallons. It was built for the USAAF in 1945 and has spent most of its life in the far north of either Canada or Alaska; it joined the Everts fleet in 1990.*

Bottom and top page 90: Air Fuels' other aircraft, **Curtiss C-46F Commando N1822M** *(c/n 22521), operates under the wonderful name of 'Salmon Ella', sporting an excellent badge under the cockpit. Both photographs were taken at Everts Air Fuels' Fairbanks base. This aircraft was built in 1945 for the USAAF and took its current registration in 1950; despite a number of different owners and operators it has retained the same one ever since. It joined the Everts fleet in 1986.*

Bottom: *Photographed at Anchorage is Everts Air Cargo* **Curtiss C-46D Commando N54514** *(c/n 33285). It is of note that the air cargo division of Everts has a full smart livery on its aircraft, but Air Fuels has just bare metal and titles. This C-46, like all the company's Commandos, has a name, in this case 'Maid in Japan'; this play on words is due to the fact that following service with the USAAF/USAF it joined the Japanese Air Self Defence Force from 1955 until its sale back to the USA in 1978. Everts bought the aircraft in 1986 and originally operated it with 'Air Cargo Express' titles but the same basic colour scheme.*

Top: *The Fairchild C-119 Flying Boxcar was first flown in November 1947 and was a development of the earlier C-82. Its role was a bulk cargo aircraft with twin booms and a high wing. Power was provided by a pair of Wright R-3350 radial pistons with an output of 3,500hp each. All new deliveries were to the military, and only at the end of their service did some appear on the civil market. A number were converted for water-bombing operators, but the severe manoeuvres needed for such operations caused airframe failures and they were withdrawn. The sole user of the type now is Flying Boxcar of Palmer, Alaska, which has a pair being brought into service. Seen here at base is* ***Fairchild C-119G Boxcar N8501W*** *(c/n 10880).*

Bottom: *With the photogenic backdrop of the snow-capped Chugach Mountain range,* ***Fairchild C-119F Boxcar N1394N*** *(c/n 10840) of Flying Boxcar is pictured at its Palmer base. This version of the C-119 has an extra Westinghouse J-34 jet fitted on the top of the cabin roof to give a boost to take-off performance. This airframe returned to Palmer from near Kodiak, where it had been parked for thirteen years following an engine problem and electrical failure while hauling salmon to a canning factory. Over the years that followed the forced-landing work took place each summer until it was ready to fly again.*

Top: *The noses of the two C-119s of Flying Boxcar are pictured together at their Palmer base in Alaska.*

Bottom: *Lockheed's Hercules has now been in production since 1954, a record for what has become the default transport for so many of the world's air forces. As well as the military C-130, the manufacturer has produced, in small numbers, a civil version under the L-100 designation. The first flight for this was in April 1964 and it stayed in the air for 25 hours! The civil aircraft do not have Funder-wing tanks or most of the purely military equipment, although they could be fitted with the wheel/ski landing system. Powerplants for the aircraft were four Allison 501 turboprops with an output of 4,050shp each. Seen at its Anchorage base is* ***Lockheed L-100-30 N401LC*** *(c/n 4606) of Lynden Air Cargo, which operates scheduled services to Bethel, Nome and Kotzebue. The aircraft can lift 48,000lb of payload in a hold 54 feet long, 10 feet wide and 9 feet high.*

Top: *Once known as Bradley Air Services, after its founder Russell Bradley, First Air calls itself the 'Airline of the North'. It operates scheduled passenger and cargo services from as far south as Ottawa to the far north of Resolute Bay, one of the coldest inhabited places on earth with a yearly average temperature of -16°C. Seen at one of the company's operational hubs, Yellowknife, is* **Lockheed L-100-30 C-GUSI** *(c/n 4600) in full First Air colours. The -30 version of the Hercules has a fuselage stretch of 6ft 8in. The aircraft can land on ice or gravel strips and has rear-opening cargo doors through which vehicles can be driven straight in, as well as items such as construction equipment for mining camps and oil exploration sites.*

Bottom: *SAFAIR of Johannesburg, South Africa, operates a mixed fleet of passenger and cargo aircraft, and specialises in leasing aircraft and contract operations including air drops and emergency airlifts. Recent operations saw the transport of five Rhinos to a game reserve in Zambia to help repopulate their numbers. Pictured at Manchester, UK, is* **Lockheed L-100-30 ZS-RSI** *(c/n 4600). Sharp-eyed readers will spot that this is the same airframe as the First Air aircraft; it was photographed prior to its sale in Canada. The South African carrier currently operates a fleet of seven L-100s with one dedicated to oil spill duties, having recently served in the Gulf of Mexico.*

Should it end like this for a once proud aircraft to be a fast food outlet? In Tulare, California, ***Convair 240/T-29 N1184G*** *(c/n 223) has had its wings clipped, literally, and has become Aerodogs – Home of the Famous Flying Wiener. Better this, though, than the scrap man's axe.*

BIBLIOGRAPHY

Gradidge, Jennifer M. *The Convairliners Story* (Air Britain, 1997)
The Douglas DC-1/DC-2/DC-3: The First Seventy Years (Air Britain, 2006)
Roach, J. R. and Eastwood, A. B. *Piston Engine Airliner Production List* (TAHS, 2002)
Propliner Magazine: various editions

AIRCRAFT

LOCATIONS

OPERATORS